BUMPOLOGY

www.transworldbooks.co.uk

BUMPOLOGY

Linda Geddes

BANTAM PRESS

LONDON · TORONTO · SYDNEY · AUCKLAND · JOHANNESBURG

TRANSWORLD PUBLISHERS
61–63 Uxbridge Road, London W5 5SA
A Random House Group Company
www.transworldbooks.co.uk

First published in Great Britain
in 2013 by Bantam Press
an imprint of Transworld Publishers

A CIP catalogue record for this book
is available from the British Library.

ISBNs 9780593069967 (hb)
9780593069974 (tpb)

Addresses for Random House Group Ltd companies outside the UK
can be found at: www.randomhouse.co.uk
The Random House Group Ltd Reg. No. 954009

The Random House Group Limited supports the Forest Stewardship Council (FSC®),
the leading international forest-certification organization. Our books carrying the
FSC label are printed on FSC®-certified paper. FSC is the only forest-certification
scheme endorsed by the leading environmental organizations, including Greenpeace.
Our paper procurement policy can be found at www.randomhouse.co.uk/environment.

Typeset in 11½/16pt Galliard by
Kestrel Data, Exeter, Devon.
Printed and bound by
CPI Group (UK) Ltd, Croydon, CR0 4YY.

2 4 6 8 10 9 7 5 3 1

To Matilda and Max

Contents

Exercise

Baby on the Brain

The Developing Baby

Section Two: Birth

Get a Move On

The Big Push

Ouch!

The Post-Pregnancy Body

Section Three: Babies

Portrait of a Newborn

Baby Bodies

Sleep

The White Stuff

Weaning

Contents

Language

The Next One

Foreword

As a newly pregnant woman, I was bursting with questions about my developing child. I remember having my twelve-week scan and spending the following weeks obsessing over what this prawn-shaped blob (which already showed signs of extraordinary wit and intelligence) was doing in there. Could it tell when I was in an aeroplane, in a swimming pool or lying down to go to sleep? Could it taste the chicken jalfrezi I was eating at the local Indian restaurant, know that it was night-time or remember the sensation of that ill-advised waltzer ride I took at the fairground?

I was also bemused by conflicting advice on eating, drinking and exercise, much of which seemed to fly in the face of common sense. Logging on to online pregnancy forums, I realized that other women were asking similar questions, often falling back on anecdote and received wisdom, which provided only unsatisfactory answers in general.

As a science journalist, I made it my job to indulge my curiosity and investigate the truth behind the old wives' tales, alarming newspaper headlines and government guidelines. So *Bumpology* was born. What started as a fourteen-part blog for the *New Scientist* during the latter weeks of my first pregnancy developed into a two-and-a-half-year obsession with the science of bumps, birth and newborn babies.

As our pregnancies developed, new and unexpected questions struck my pregnant friends and me, and as 'B-day' approached and our thoughts (and worries) turned to the process of birth itself, I found myself asking even more.

The journey didn't stop there. Once our daughter, Matilda, was born, a whole new world of experiences and unresolved queries tugged at my curious mind. In the darkest hours of sleep deprivation, I wondered if babies could tell the difference between night and day and whether the personality traits babies seem to be born with would follow them into later life; when my baby's nappy exploded, I considered the sterility of baby poo; and as our daughter grew more mobile and started to develop language, I contemplated just when and how qualities such as long-term memory and empathy develop.

As it turned out, many of the answers were already out there; buried in scientific journals or lodged in the brains of academics. It just took some digging to find them.

Shortly after starting to write this book, I fell pregnant with my second child, Max. This time it was a very different experience. The more I researched, the more I realized that much of what I had been told during my first pregnancy was not backed up by evidence, and, in some cases – such as being told that if I requested epidural pain relief I was more likely to need a C-section – was plain wrong.

I became increasingly angry. Every week, expectant parents are given new things to worry about. Pregnant women mustn't eat too much as it may raise the baby's risk of obesity or diabetes, but we mustn't diet as that could have a similar effect. Neither can we exercise, for fear of triggering a miscarriage. It's enough to raise your blood pressure just thinking about it; only we mustn't get stressed because that's bad for the baby too. And if we do get stressed, we can't drink alcohol, go for a spa treatment or lie on our backs to relax.

It doesn't get much better once the baby is born. Before having a first child, most of us have never been left alone with a newborn, let alone been handed sole responsibility for its survival and well-being. Some of us get through it on instinct, others consult every baby book going – but even these offer conflicting advice on key topics like sleep, feeding and crying. Most books are written by people with years of experience, but had anyone systematically compared the techniques they recommend and figured out if any one were better than the other – or if they may be harmful to some babies? I wanted to find out.

Bumpology is my attempt to make sense of all the conflicting advice that's out there about pregnancy, birth and raising babies. I am not a doctor, and so none of what I write is intended as medical advice – although you may find it useful to refer doctors and midwives to some of the studies I cite in the Notes section. What I do have experience of is wading through the scientific research and asking questions of doctors and scientists, in an attempt to make sense of it all.

The first thing to understand is that many of the newspaper headlines we read are based on one study, often preliminary, sometimes conducted on animals. Although it's in the nature of journalism to simplify, sometimes this has the unfortunate consequence that readers take what journalists write (or how their editors choose to spin it) too seriously. My husband – a fellow science journalist – has an anecdote about when he went to see a nurse to have his sperm count tested for a feature he was writing. He was told that he should avoid using his laptop on his lap, as the heat this generates could damage the quality of his sperm. When he asked what the source of this advice was, his own story in the *Daily Telegraph* was quoted back at him. It was based on preliminary research that was presented at a scientific conference on a slow news day. Although he had written some of these caveats into the original article, they had been lost in translation.

Then there are times when advice is offered for the benefit of a population, rather than individuals within that population. This is particularly true of guidance issued by global bodies such as the World Health Organization which are trying to do their best by women in a whole host of different economic and social situations. Telling women that they need to breastfeed for two years may save the lives of many infants in developing countries through the protection it affords against infectious disease – but it also risks causing unnecessary anxiety among women in Western countries who perhaps can't breastfeed or who stop after a couple of months because they have to return to work. Advice also tends to be aimed at the average individual, without taking the specific needs and circumstances of an individual into account.

It is important to understand that science doesn't yet have a definitive answer to many questions, even though plenty of research may have been done. Take the question of how much alcohol it's safe to drink during pregnancy. We know that there are defined risks above a certain threshold, but below this there's a grey zone where the evidence may be mixed or conflicting. In other cases such as eating peanuts, taking a jacuzzi, or exercising, there may be a theoretical risk, but there's very little good-quality evidence to support it.

Faced with such uncertainty, many doctors and health

organizations will err on the side of caution, hence the long lists of things that pregnant women should avoid. You can't blame them; they don't want to give the wrong advice and risk someone falling ill (or worse) as a result. Yet it is my strong belief that women (and men) should be equipped with the facts they need in order to be able to weigh up risks and make decisions for themselves.

So how should one go about making sense of alarming newspaper headlines and statistics such as 'First-time mums are three times more likely to lose their baby if they give birth at home'?

One thing to look out for is the size of the study. If it has been conducted with fewer than a hundred people, I'd be tempted to regard the results as provisional at best; if with hundreds or thousands of people, it may be more reliable – although it also depends on what is being measured. If it's a relatively rare event such as death during childbirth, then, ideally, you'd want to see a study of many thousands of people, as this should increase your confidence that the finding isn't just down to chance.

Unfortunately, many of the studies I examined during the research for *Bumpology* fell into the tens to hundreds of people category. In some cases, this was the only evidence that existed, so it was all that I had to go on, and I have tried to make this clear to the reader. In other cases, plenty of small studies had been done, but they reached conflicting conclusions; here, I have generally turned to systematic reviews in which the results of lots of studies are pooled, to paint as broad a picture of the evidence as possible. The highly respected Cochrane Collaboration is an international network of researchers which specializes in doing just that, and it has reviewed many of the key topics, particularly in the area of childbirth.

Finally, some words about risk and statistics. Often, newspapers will quote research that found an x per cent increased risk of something (generally nasty) happening. What they're usually referring to is the relative risk, which tends to paint a far scarier picture than the figure we really care about – the absolute risk. For example, you may be told that women who request epidural pain relief during labour are 42 per cent more likely to need additional help delivering their babies (through the use of forceps or a vacuum device called a ventouse). However, the risk of needing an instrumental delivery among the general population is pretty low (around 12 per cent), so what we're actually talking about is an increase of 42 per cent of that 12 per cent, i.e. only a 5 per cent increased risk. Another way to look at it is that for every twenty women who request epidural pain relief, there will be one extra instrumental delivery.

When weighing up risks, it's sensible to ask yourself a couple of questions. Firstly, what is the risk of the bad thing happening to start with, and just how serious are the consequences if it does happen? Secondly, how much does the action you're considering increase that risk? Finally, how great is the benefit if you decide to do it anyway?

In writing this book, I have tried to cut through some of the confusion and conflicting advice, leaving parents with the facts about the state of the scientific evidence. It is for individual parents to decide for themselves how much risk they want to take, and what is going to work best in their individual circumstances. Having a baby can be one of the greatest joys that life bestows, but it is hard work. We can do without any unnecessary guilt, anxiety and doubt.

It's also a time of great wonder. I still find it staggering that a single egg and sperm can meet and trigger this cascade of events that lead to a new little person being built and then unfolding into this fascinating, lovable and endlessly entertaining individual. In researching this book I've learned things about babies that will never cease to amaze me – indeed, I may never look at my children in the same way again.

Bump

Food and Drink

1

Why do pregnant women crave unhealthy food?

ANGELINA JOLIE CRAVED chocolate with cinnamon and chilli, Cate Blanchett wanted pickles and ice cream, while Britney Spears is said to have wanted to eat soil. A recent study on the pregnancy website www.babycenter.com found that about 85 per cent of American women experienced at least one food craving during pregnancy. Around 40 per cent craved sweet foods, while 33 per cent lusted after salty snacks. Spicy foods came in third, at 17 per cent, followed by fresh fruit, at 10 per cent.

Surprisingly few scientific studies have been carried out on food cravings during pregnancy. Those that have been find that pregnant women often have a desire for salty foods and become less sensitive to the taste, as well as becoming more sensitive to bitter tastes.

The salt craving may come about because, as their blood volume expands, pregnant women need slightly more salt to maintain the balance of fluids in their bodies. An aversion to bitter flavours is less easy to explain, but some have proposed that it steers women away from poisonous plants, which are often bitter (see 3: 'What causes morning sickness?').

As to why pregnant women crave sweet foods, studies of how taste preferences vary over the menstrual cycle may provide some clues. Women tend to want foods that are high in carbohydrate and fat during the second half of the menstrual cycle, when levels of the hormone progesterone are high. They are more sensitive to sweet tastes during the first half of the cycle, when levels of oestrogen are higher.

Although both hormones increase during pregnancy, there is relatively more progesterone than oestrogen, which may explain why women long for sweet, energy-rich foods like chocolate or cake (or preferably both at the same time).

A lot of people assume that you crave the foods your body needs. Although no one has specifically studied this in pregnant women, there is some evidence that it may be true. Leigh Gibson and his colleagues at the the University of Roehampton, London, gave their test group two flavours of soup to try, one that was high in protein and one that was low. A few days later, the volunteers were either starved of protein or given a protein-rich drink before being offered a choice of the two soups for lunch. Those that were protein-starved said they preferred the protein-rich flavour, and they also ate far more of it. In other words, if you are starved of protein, you subconsciously begin to prefer the taste of foods that contain large amounts of it. 'It seems we can rapidly learn to want to eat foods that supply needed nutrients like protein,' says Gibson.

Likewise, cravings for sweet, fatty foods may reflect a need for energy, although Gibson cautions that they could also reflect a need for emotional comfort, as we often get a lot of pleasure from eating them. So if your body is screaming out for ice cream, perhaps it's worth considering how much you've already eaten that day before concluding that your baby is deficient in chocolate-chip cookie dough.

2

Do pregnant women really eat coal?

'HAVE YOU BEEN eating coal?' was the first thing my dad asked me upon learning I was pregnant. Strange as it sounds, some women develop a taste for minerals, metallic objects and even soil during pregnancy, a phenomenon known as 'pica'. Some have proposed that it's a sign of iron deficiency – though how much iron you'd get from a mouthful of coal is questionable.

A survey of 2,231 British women found that 31 per cent claimed to have experienced unusual cravings during pregnancy. Top of the list was ice, followed by coal, toothpaste, sponges, mud, chalk, laundry soap, matches and rubber. One of my neighbours even reported a penchant for cigarette ash during her pregnancy.

Many of these phenomena have their own names. Pagophagia, or ice-eating, is also common among American women: a separate survey found that 18 per cent of women in the state of Georgia did it, the amount they ate ranging from a couple of glasses of ice cubes to several kilograms a day.

Earth-eating, or geophagia, is particularly common in African countries such as Tanzania, where up to 60 per cent of pregnant women indulge. Clay-rich soil seems particularly desirable, although pottery is sometimes eaten instead.

Scientists have proposed various explanations for pica,

including cultural trends, stress relief (like biting one's fingernails) and hunger. Some women who have been interviewed about their pica say that it helps to relieve heartburn and nausea, and often the items they choose have an alkaline pH, which may help to neutralize stomach acid. Most women, though, have no idea why they crave such oddities.

A common scientific explanation is that these women are deficient in certain nutrients, such as iron, zinc or calcium. Although there have been hundreds of reports of iron-deficient women craving soil and other pica, when researchers have analysed how much iron is released from the digestion of soil or clay, the answer is: very little. Neither does giving women iron supplements to cure their anaemia stop their cravings. What's more, eating clay and soil can actually inhibit the uptake of iron and other minerals by the gut. But just because soil doesn't cure anaemia, it doesn't mean anaemia doesn't trigger the craving. 'It may be that being deficient in essential minerals induces a craving for anything mineral-tasting,' says Gibson.

An intriguing suggestion is that eating clay reduces the chances of food poisoning. Several studies have shown that certain substances in soil can bind to bacteria, viruses and toxins in the gut, preventing them from being absorbed into the bloodstream. Some types of clay also seem to increase the secretion of mucus in the intestines, making it harder for bacteria to get through.

Why women should crave ice remains a mystery. Some have suggested that it has the same crunchy texture as dry clay or soil, so eating it has a similar psychological effect. Others propose that it calms tongue pain or swelling, which is a common symptom of iron deficiency.

3

What causes morning sickness?

I WRITE THIS in the throes of morning sickness, seven weeks pregnant and curled up in bed with a hot-water bottle, praying for the current wave of nausea to pass. This is my second pregnancy, and I'm sure it's worse this time (for one thing, I'm actually vomiting, rather than just feeling queasy).

Up to 80 per cent of women experience some degree of morning sickness during pregnancy. Generally, it is at its worst between weeks four and ten, and even the unlucky few who continue to feel sick after ten weeks are usually free of it by week twenty.

There are several theories about why morning sickness should exist. One is that it evolved to steer pregnant women away from eating plants containing toxins that might harm the baby (also why pregnant women are thought to develop an aversion to bitter foods; see 1: 'Why do pregnant women crave unhealthy food?'). But, in that case, you'd expect women to feel sicker after they'd eaten fruit or vegetables, and there's little evidence to support this – although in my first pregnancy I did develop a weird aversion to lettuce. This theory is also dealt a minor blow by the fact that India (where many people are vegetarian) has the lowest rate of morning sickness, with just 35 per cent of women suffering from it. Japan has the highest rates, with 84 per cent of pregnant women experiencing nausea and vomiting.

A related theory is that morning sickness evolved to protect the embryo from bacteria or parasites in the diet that could harm it. Paul Sherman and Samuel Flaxman at Cornell University in New York put forward this theory after reviewing studies involving nearly 80,000 pregnant women from around the world. Pregnant women are particularly susceptible to food poisoning because their immune systems are suppressed, and serious illness can increase the risk of stillbirth or miscarriage (one reason why we're told to avoid soft cheeses). Cells in the embryo are also dividing rapidly, leaving them more susceptible to any damage to DNA caused by harmful toxins.

Sherman and Flaxman note that the level of nausea and vomiting usually peaks at the same time as the baby's vital organs are forming, and they are also growing very quickly at this point, so if their DNA is damaged it could have disastrous effects. Commonly avoided foods include meat, fish, poultry and eggs (foods that are the most likely to contain dangerous micro-organisms), as well as caffeinated drinks like tea or coffee which could harm the baby if consumed in large amounts.

It's true that morning sickness is less common in societies that consume only small amounts of meat and fish and eat more plants as staple foods – including some American Indian and First Nations tribes. Women experiencing morning sickness may also be marginally less likely to miscarry. One large study found that 90 per cent of women with no morning sickness gave birth to a live baby, compared to 96 per cent of women who did experience morning sickness.

Perhaps more convincingly, morning sickness may simply be an unfortunate side effect of the same hormone that confirmed you were pregnant in the first place. Human chorionic gonado-trophin (HCG), the hormone detected by pregnancy tests, is produced by the embryo during the early weeks of pregnancy in order to maintain a structure called the corpus luteum, which produces other pregnancy hormones needed to keep the lining of the uterus nice and thick until the placenta is big enough to take over.

But HCG also has other effects on the body, such as stimulating the thyroid gland in the neck. This gland produces several hormones that regulate metabolism, and high levels of thyroid hormones seem to be associated with particularly severe vomiting in women with morning sickness. Two other pregnancy hormones, progesterone and oestrogen, may also contribute to general feelings of queasiness, bloating and heartburn, as they slow down the digestive system, so food takes longer to travel through your gut. Rampaging progesterone may explain another annoying side effect of early pregnancy: extreme tiredness. At very high doses, progesterone can be used as an anaesthetic, having a similar effect on the brain to barbiturates.

No one has conclusively shown that women with higher levels of these hormones are more prone to morning sickness, it's just a theory. Neither do we know why some babies seem to cause more sickness than others – although the placentas of female foetuses do seem to produce more HCG than those of male foetuses (presumably because hormone production is linked to the X chromosome). Just three weeks into pregnancy, HCG levels are nearly a fifth higher in women carrying female embryos than those carrying males. There is also some evidence that women with particularly severe morning sickness are marginally more likely to be carrying girls or twins.

4

Is there anything I can do to reduce morning sickness?

PERHAPS, ALTHOUGH THE evidence isn't very strong. Both the American College of Obstetrics and Gynecology (ACOG) and the UK's National Institute for Health and Clinical Excellence (NICE), which offers evidence-based guidance to doctors, suggest that ginger may help to relieve nausea and vomiting. ACOG also recommends vitamin B6, or vitamin B6 plus an antihistamine called doxylamine.

However, the Cochrane Collaboration recently reviewed the evidence for morning-sickness treatments including acupuncture, acupressure, vitamin B6, ginger and some anti-sickness drugs and found little consistent or reliable evidence to support any of them. Instead, you may just have to ride out the sickness and look forward to the later stages of pregnancy, when you'll be carrying a proud bump and should be able to feel your baby moving about inside you.

5

How much alcohol is it safe to drink during pregnancy?

I'M SURE I'M not alone in feeling confused, and a little irritated, by the messages women are given about alcohol consumption during pregnancy. A recent discussion on the UK internet forum

Mumsnet about whether someone could indulge in a few glasses of wine at a wedding generated sixty-nine responses – with angry arguments on both sides.

A withering list of countries, including the UK, US, France, Canada, Australia, New Zealand, Ireland, the Netherlands and Spain, currently advise pregnant women to abstain completely from alcohol during pregnancy (although the UK government concedes that if you do decide to drink, you should 'consume no more than one to two units once or twice a week'). The US advises women who are even thinking about getting pregnant to abstain.

Yet, until quite recently, pregnant women were told to drink half a pint of Guinness or stout each night to keep their iron levels up. We all know of women who have drunk moderate amounts of alcohol during pregnancy – or who have been blind drunk before realizing they were pregnant – and their babies appear to have turned out fine.

So what is a safe amount of alcohol to drink? When researchers at the UK's National Perinatal Epidemiology Unit in Oxford reviewed forty-six scientific papers looking at the effects of low to moderate drinking during pregnancy, they found no increased risk of miscarriage, stillbirth, premature birth, low birth weight or birth defects (including foetal alcohol syndrome). These women were drinking a fair amount of alcohol each week – up to 10.4 units per week, or about five small glasses of wine. However, the researchers also found plenty of flaws in the studies, which meant they couldn't say conclusively that light to moderate drinking is safe.

The truth is that no one really knows what constitutes a 'safe' amount of alcohol. It's clear that large amounts can cause foetal alcohol syndrome (FAS), a serious condition that stunts the physical and mental development of the child. According to a report by the UK's Royal College of Obstetricians and Gynaecologists, drinking more than six units a day puts babies at risk of FAS, and can also trigger miscarriage. Birth defects are three times more common in the babies of women who drink more than four and a half units a day, compared to light drinkers and abstainers – but birth defects are still pretty rare. Meanwhile, 45 per cent of women who drink more than two units a day (a small glass of wine) during their third trimester may have a baby with low birth weight, which can increase the risk of developmental or behavioural problems later in life – although the evidence for this is mixed.

However, below this level there's this big grey zone in which there may or may not be an effect. In 2012, a large Danish study concluded that one to eight drinks a week seemed to have had no effect on children's IQ, attention span or on brain functions such as planning, organization and self-control by the time they reached the age of five. Unlike many, the study was well designed, in that it asked women how much alcohol they were drinking when they were actually pregnant, and then followed the progress of their children as they grew older.

It's also not correct to assume that abstinence is the best policy: several studies have now suggested that children born to mothers who drank one to two units of alcohol per week in their pregnancy were less likely to have behavioural problems than those born to abstainers. In one such study, the sons of light drinkers had fewer behavioural troubles and less hyperactivity, and the daughters had fewer peer problems and emotional difficulties. Children of moderate drinkers, who consumed three to six units per week, were broadly similar to those of women who drank nothing, while there were slightly more behavioural problems in the children of women who drank more than seven units a week. However, it's also possible that these differences arose because light drinkers were more relaxed, or came from higher income or more educated backgrounds than abstainers.

Patrick O'Brien of the RCOG says that light drinking – one to two units once or twice a week – is extremely unlikely to harm your baby. Vivian Nathanson of the British Medical Association (which advocates abstinence) admits that up to eight units is unlikely to cause harm, but suggests it is far better to err on the side of caution and drink nothing, or very little.

There may, however, be other reasons why women are told to abstain. One is that it's very difficult to do broad enough studies on light drinkers to catch the very small number of problems that might crop up. It's possible that light drinking does carry a small risk, but we just don't know about it yet. Another is that women are not trusted to be capable of following guidelines on a recommended number of units. 'We know that a lot of people drink a great deal more than they think they do, because what they think is a unit is in fact several units,' says Nathanson. 'Abstinence is an easy thing to measure.'

It's also likely that babies are more vulnerable to the effects of alcohol at some periods of their development than at others. One recent study found that drinking more than one alcoholic drink per day during pregnancy generally increased the risk of babies being underweight or shorter in length at birth, but consuming alcohol between weeks seven and twelve was strongly linked with the characteristic deformities associated with foetal alcohol syndrome: a small head, small upper jaw, thin upper lip and small, narrow eyes.

Women who just had a couple of sips of alcohol were classified as non-drinkers, however. Weeks seven to twelve of pregnancy are particularly busy in terms of development: in just five weeks, the baby goes from looking like a mutant tadpole to looking like a human being, while after week twelve its main task is simply to get bigger. So perhaps it's no surprise that this would be a particularly critical time to avoid anything that might perturb that development. It's also probably the time when women are least likely to want to drink, however, as morning sickness is at its peak and most women will be feeling dog-tired.

Ultimately, every woman needs to weigh up for herself whether she believes the occasional drink during pregnancy is OK. But, based on the evidence I've seen, I've developed my own set of rules.

- The first trimester is when the baby is most at risk, so try to avoid drinking then. If you didn't realize you were pregnant, just stop drinking now. Most women realize they are pregnant by seven weeks, which is when the most critical point seems to start.
- If abstinence isn't for you, try to stick to one to two units once or twice a week. However, if you occasionally drink more than this (up to about eight units a week), it's probably fine.
- Know how much a unit of alcohol is, and remember that some wines and beers are stronger than others.
- Binge drinking is likely to be worse than drinking the same amount over a long period because it exposes your baby to higher concentrations of alcohol. So try to spread your drinks out as much as possible.

6

Can unborn babies taste what Mum is eating?

UNBORN BABIES ARE floating in a veritable cocktail of taste sensations, which may be teaching them what's good to eat in preparation for when they pop out into the big wide world.

Taste buds on the tongue begin to develop just thirteen to fifteen weeks into pregnancy, enabling babies to detect simple tastes such as sweet, sour and salty, but many of the more complex flavours we experience are the result of volatile molecules from foods like garlic passing over smell receptors in the nose.

All of these things can find their way into the amniotic fluid that surrounds the baby, in much the same way that flavours also get into breast milk (see 111: 'Does what I eat change the flavour of my milk?'). 'If it gets into the blood supply, it will get into the amniotic fluid and the breast milk,' says Julie Mennella at the Monell Chemical Senses Center in Philadelphia, Pennsylvania, who spends her days trying to understand how preferences for certain flavours develop.

During the third trimester of pregnancy, foetuses breathe and swallow around a litre of amniotic fluid a day – probably as practice for when they're born, rather than for any nutrients – and much of this fluid will pass smell and taste receptors in the nose and mouth.

Some smelly molecules, such as the sulphurous compounds that give garlic its characteristic smell and taste, are so potent that if you take a sample of amniotic fluid from a woman who has eaten a garlicky meal, you can smell the garlic wafting from it.

So unborn babies do get a taste of what their mums are eating during pregnancy, and they also seem to remember it. Several studies have shown that babies whose mothers ate a lot of garlic or aniseed during pregnancy are attracted to those smells in the first days after birth, making rooting or sucking motions when they catch their scent.

7

Can Mum's food fads influence her baby's palate?

NOT ONLY ARE babies treated to a banquet of flavours before birth, it seems the memory of these tastes may influence the kinds of foods they like in later life.

Studies have shown that if Mum drinks a glass of carrot juice four times a week during the last trimester of pregnancy or the first two months of breastfeeding, her baby will find cereal prepared with carrot juice particularly appealing once it has been weaned. Babies whose mums eat a lot of fruit during pregnancy are also more likely to enjoy fruit during weaning. 'The baby is learning what foods Mum likes,' says Mennella. 'I think it's the first way they learn what foods are safe and also what foods are available.'

Trying to eat as many different flavours as possible during pregnancy and breastfeeding should therefore expose your baby to a smorgasbord of different tastes, and in theory this may make them less prone to fussy eating. However, it could also set a bad example if you choose to indulge in less wholesome options. Babies whose mothers drank two small glasses of wine a week during pregnancy showed more smiling, suckling and licking expressions in response to the smell of alcohol than those whose mums drank very little or nothing at all – although it's not clear whether this penchant for alcohol lasted into adulthood.

A bizarre study on rats suggests it might. Adolescent rats whose mothers had consumed alcohol during pregnancy showed a greater interest in drunken partners than those whose mothers hadn't consumed alcohol – presumably because they could smell the alcohol on their breath.

It gets even weirder. Several studies have found that children's propensity for salt can be influenced by the degree of morning sickness their mothers experienced. Sixteen-week-old babies whose mothers endured moderate to severe vomiting during the first fourteen weeks of pregnancy were more tolerant of salty drinks than those whose mothers experienced no morning sickness. One possible explanation is that throwing up can make women de-hydrated, making their amniotic fluid more concentrated and salty.

8

Is coffee bad for my baby?

IF YOU'RE SOMEONE who simply can't function without their morning latte or espresso, the great news is you don't have to stop. Although a handful of studies have linked the consumption of caffeine to reduced birth-weight and an increased risk of preterm birth or stillbirth, other studies have found no effect.

The Cochrane Collaboration tries to make sense of exactly these kinds of contradictory studies, by taking a bunch of them and systematically weighing up the evidence, to help people make informed decisions about their health. It recently weighed in on the issue of caffeine during pregnancy and concluded that drinking three cups of instant caffeinated coffee a day during early pregnancy probably has no effect, so there is no reason to avoid it in small amounts.

Drinking a lot more than three cups of coffee a day may be more of a problem, however. One large study found that women who drank more than 550mg of caffeine a day (roughly equivalent to six regular cups of coffee, or twelve cups of tea or green tea) had slightly shorter babies.

If you do decide to drink coffee, one thing to watch out for is that different blends vary in the amount of caffeine they contain. In the UK, pregnant women are advised to limit themselves to 200mg of caffeine a day – somewhere between two and four cups of coffee, depending on whose data you use. But a recent survey of espressos bought in Scotland found a sixfold variation in the amount of caffeine they contained, with one espresso bought in Glasgow's West End containing 322mg. A Starbucks espresso, by contrast, contained just 51mg. Since lattes and cappuccinos are often made by diluting a double shot of espresso, you might also want to ask your barista to hold the second shot.

As a rough guide, here's what the UK's Food Standards Agency says about how much caffeine common food and drinks contain:

- One mug of instant coffee: 100mg caffeine (although other studies suggest it may be as low as 50mg – I guess it depends on how strong you like your brew)
- One mug of filter coffee: 140mg caffeine
- One mug of tea: 75mg caffeine
- One can of cola: 40mg caffeine

- One can of energy drink: 80mg caffeine
- One 50g bar of plain (dark) chocolate: 50mg caffeine
- One 50g bar of milk chocolate: 25mg caffeine

9

Can I eat peanuts during pregnancy?

THE OFFICIAL ADVICE on this waxes and wanes, but, currently, both the UK Food Standards Agency and the American Academy of Pediatrics (AAP) say that it's fine to eat peanuts during pregnancy (even though both previously advised against it if there was a history of allergy or asthma in the child's immediate family).

Certainly, several studies have suggested an indirect link between eating peanuts during pregnancy and childhood allergy. For example, in 2010, many newspapers seized upon an American study which found that children were more likely to test positive for antibodies against peanuts if their mothers had eaten peanut products during pregnancy (an indirect marker for possible peanut allergy). However, even the study's authors urged caution in interpreting its results. For one thing, they relied on mothers' memories of how many peanuts they had eaten, rather than measuring quantities themselves. They also tested only children who already had suspected allergies to milk or eggs rather than the general population, and they didn't test for peanut allergy directly.

What's more, several studies have found the opposite: that eating peanuts during pregnancy may protect infants against allergies. In such situations, the most sensible thing to do is to weigh up the results of as many studies as possible to decide where the best evidence lies. Fortunately, someone has done this for us: in 2008, the UK's Committee on Toxicology reviewed a slew of human and animal studies and concluded that there aren't enough high-quality studies to support the idea that eating peanuts during pregnancy increases your child's risk of allergy. A recent review by the American Academy of Pediatrics reached a similar conclusion. This doesn't necessarily mean peanuts have the all-clear; just that more good-quality research is needed to decide either way. Until this gets done, you may as well carry on eating peanut butter – spread on gherkins, if you fancy it.

10

Should pregnant women really eat for two?

THERE'S NO DOUBT that being pregnant is hungry work and, no, you're not just being greedy: the hormone progesterone, which soars during pregnancy, is known to boost the appetite. But although it can be tempting to help yourself to a second slice of cake because you're 'eating for two', there are good reasons not to overindulge – at least not every day. The good news is that if you can't resist, it is generally safe to exercise during pregnancy (see 22: 'Is it safe to exercise during pregnancy?').

In 2009, the US Institute of Medicine issued guidelines saying that women should gain no more than 11.5–16kg (25–35lb) during pregnancy, while those who are already overweight should gain no more than 7–11.5kg (15–25lb), and obese women no more than 5–9kg (11–20lb).

That's all very well, but around half of women gain more weight than this, and although a little extra weight is unlikely to harm you or your baby, putting on a lot of extra weight can increase your risk of gestational diabetes, high blood pressure, and needing a C-section or mechanical help to get the baby out.

Another good reason not to overindulge is that women who put on a lot of extra weight during pregnancy find it harder to shift it afterwards – and I speak from bitter experience. A recent review of nine studies found that women who gained more than the recommended amount of weight carried an average extra 3.06kg three years after giving birth and an extra 4.72kg after fifteen years. Putting on too much weight too quickly may also increase your chances of having haemorrhoids, varicose veins and stretch marks.

So how many extra calories does your body need to grow a baby? Most doctors recommend eating just 100–300 extra calories a day during pregnancy, which doesn't even add up to a slice of cake.

As a rough guide, you should expect to put on 1–2kg (2–4lb) in total during the first twelve weeks, and then 0.5–1kg (1–2lb) a week after that.

11

How dangerous is it to eat Camembert and blue cheese?

WHY IS IT THAT, when you're pregnant, beautiful creamy soft cheese is everywhere? And yet we can't eat it – we're told – because soft and blue-veined varieties of cheese carry a risk of listeriosis, a type of food poisoning that can trigger birth defects and miscarriage.

Listeriosis is caused by a bacterium called Listeria mono-cytogenes which is present in soil and therefore gets on to the grass that the cows, sheep and goats which produce the milk for cheese-making eat. These bacteria are usually destroyed by heating or pasteurization but, if they do survive, they find it easier to grow in ripened soft and blue-veined cheeses, which are often less acidic and contain more moisture than hard cheeses.

Although people often come into contact with listeria without getting listeriosis, pregnant women are twenty times more likely to develop it than non-pregnant women of the same age because their immune systems are weakened. Even so, cases of listeriosis are pretty rare. According to the UK's Food Standards Agency, there are around 230 recorded cases in the UK per year, while the Centers for Disease Control and Prevention (CDC) in the US says there are around 1,600 cases there a year (resulting in around 500 deaths), and only around 17 per cent of these cases are in pregnant women. People over the age of sixty are also at increased risk, as are people with cancer or any other disease that suppresses the immune system.

That said, the consequences of listeria infection during pregnancy are pretty terrifying. Around one in 25,000 pregnant women contracts listeriosis, and, besides feeling rotten themselves, the infection can cause complications and trigger miscarriage in around 20 per cent of cases.

Unfortunately, we can't say how many cases of listeriosis there would be if pregnant women decided to ignore the standard advice and scoff artisan cheeses at will. But cheese isn't the only source of listeria infection – pre-packed sandwiches, ready-to-eat meats such as hot dogs, sliced meat and pâté, butter and smoked salmon can also carry the bug, so if you're worried about listeria, you should probably avoid these as well. A recent analysis of the likely sources of listeria infection in the US found that melons and hummus were prime suspects, as was living on a cattle farm, while a review of recent UK cases found that foods such as pre-packed sandwiches or ready-to-eat salads topped the list of likely sources of infection.

The good news is that thoroughly cooking risky foods should kill the bacteria that cause listeriosis – as long as you ensure the food is piping hot before serving. There are mixed messages about whether soft cheese made from pasteurized milk is safe. The CDC says it is, while the NHS cautions against eating soft cheese, regardless of whether it says it has been pasteurized. Pasteurization will kill any bugs that were living in the milk before it was treated, so should certainly reduce the risk, but there is still a chance that cheese could become contaminated during packaging.

The Pregnant Body

12

Can the shape of my bump or anything else predict the gender of my child?

A BELLY LIKE a basketball, bad morning sickness, weird food cravings . . . If you believe the old wives' tales, if you have these you're definitely expecting a boy. But are any of these guessing games underpinned by scientific evidence?

Let's take a look.

Women who have basketball-shaped bumps are expecting boys

Women who carry high and all upfront are expecting boys, while a low, wide bump is a sign of a girl – or so says the folklore. But Janet DiPietro of Johns Hopkins University has investigated and found no correlation between the shape of a woman's bump and the sex of the child that is developing inside it.

She asked 104 pregnant women what gender of child they thought they were carrying, and whether they thought their bump was all upfront or spread around the hips. The interviewer also made an assessment of this. There was only moderate agreement between the two, suggesting that the shape of a woman's bump is pretty subjective, and neither assessment correctly predicted the gender of the child that eventually popped out.

Bad morning sickness means you're having a boy

Actually, the opposite is true: mums-to-be with really bad morning sickness are more likely to have a girl. In 1999, researchers analysed the records of 8,186 Swedish women whose morning sickness was so bad that they had to be admitted into hospital and found that 44.3 per cent of them gave birth to boys (compared to 51.4 per cent of the general population).

They blame higher levels of the pregnancy hormone HCG in women carrying girls. This is the hormone that pregnancy tests detect, and is produced at the beginning of pregnancy in order to maintain a structure called the corpus luteum which churns out other pregnancy hormones until the placenta is developed enough to take over. No one really understands why female foetuses are associated with higher levels of HCG, but differences have been detected as early as sixteen days after fertilization and continue throughout pregnancy. HCG may be responsible for keeping pregnancies going, but it also comes with some side effects. It can make your bladder more over-reactive and make you feel sick.

Shortly after this study was published in the *Lancet*, Henrik Toft Sørensen of Aarhus and Aalborg University Hospitals in Denmark wrote to the journal, claiming that marital status also influenced the sex of the baby, with fewer boys born to single mothers. Like the Swedish group, he found that women with severe morning sickness were more likely to give birth to girls, but the correlation was even stronger in single mothers. Among women with severe morning sickness, 40 per cent who lived alone had boys, compared with 45 per cent who lived with their partner (and 51 per cent of women who didn't experience morning sickness).

Foetal heart rate predicts the baby's sex

One of the most exciting moments of pregnancy is when your doctor holds a Doppler probe to your belly and you hear your baby's heartbeat for the first time. Because foetal hearts beat much faster than adult hearts, this often sounds like galloping horses.

According to this old wives' tale, if your baby's heart rate is faster than 140 beats per minute (bpm) you're expecting a girl, and if it's slower, you're expecting a boy. The trouble is that the heart rate tends to decrease during pregnancy, from 170–200bpm at eight to ten weeks down to 120–160bpm by mid-pregnancy, and there seems to be little difference between boys and girls – at least during early pregnancy.

In one study, researchers used ultrasound to measure heart rate in 477 foetuses before the fourteenth week of pregnancy. The average heart rate of girls was 151.7bpm, and for boys 154.9bpm, but this wasn't a big enough difference to be statistically significant and would be too unreliable to use to guess the baby's gender.

Interestingly, a bigger difference can be detected during labour itself, when female babies do seem to have faster heart rates. The reason for this is unknown.

Strange food cravings mean you're having a boy

There's not much evidence to support this one, but one study did find that women who are carrying boys have bigger appetites. It recorded the eating habits of 244 pregnant women and found that those who gave birth to boys ate 10 per cent more calories per day than those who had girls. The trouble is: how do you know if you're hungrier than the next pregnant mum? Indeed, the authors of this study admitted that the difference was not big enough to predict the sex of a baby with any accuracy.

Women 'just know'

'Female intuition' claims power over many things – including being able to predict the sex of unborn children. And it may be one of the best tools going for predicting your baby's gender. Victor Shamas at the University of Tucson in Arizona asked 108 pregnant women to guess the sex of their baby. Seventy-five of them claimed to have a gut feeling about it or had experienced a dream, and 60 per cent of them were correct – which is more than you'd expect by chance. When the researchers discounted women who had a strong preference for a boy or a girl, the forty-eight remaining women guessed right 71 per cent of the time. Unfortunately, the study was never published, but it does provide ammunition for women whose partners keep on suggesting ridiculous boys' names: 'Sorry, darling, but I just know it's going to be a girl . . .'

If the egg implants on the right side of the uterus, it's a boy; if on the left, it's a girl

This is not so much an old wives' tale as an interesting piece of research posted on an internet forum for obstetricians in 2007 which has generated a lot of discussion. A Canadian sonographer called Saad Ramzi Ismail claimed to have scanned 5,376 pregnant women and found that the position in which the placenta attaches to the uterus strongly predicts whether a woman is having a boy or a girl. He said that 97.2 per cent of male foetuses attach on the right side of the uterus, while 97.5 per cent of female foetuses attach on the left. Although Ramzi's work was written up as a scientific paper, it doesn't appear to have been published in any peer-reviewed journal, so his methods haven't been subjected to any detailed scientific scrutiny. Neither did he give any satisfactory explanation as to why the location of the placenta should influence the baby's gender. One earlier study by an American group suggested that women with low-lying placentas (a condition called placenta praevia) were 14 per cent more likely to have boys, although the biological mechanism behind this is also unclear. One possibility is that the time of conception might influence where in the uterus an embryo implants (as there's some evidence that boys are more likely to be conceived if you have sex at the time of ovulation: see 149: 'Is there anything I can do to influence the gender of my baby?'). Alternatively, temperature differences in different areas of the uterus might better favour the survival of male or female embryos. In 2010, an Australian group announced at the World Congress on Ultrasound in Obstetrics and Gynecology that they had tried to replicate Ramzi's findings. They recorded placental location in 277 women, and found no convincing evidence that it correlated with the sex of the baby.

13

Why don't pregnant women topple over?

EVERY YEAR, a magazine called the *Annals of Improbable Research* awards the 'Ig Nobels' – prizes that celebrate the unusual, honour the imaginative and spur people's interest in science, medicine and technology. In October 2009, the physics prize went to a group of anthropologists who dared to ask the question: 'Why don't pregnant women topple over?'

As a woman's body grows during pregnancy and her bump begins to stick out, this changes her centre of gravity, making her more clumsy and prone to waddling. But it's rare to see a pregnant woman topple over – even on London's Tube system, where few people are willing to stand up and offer her a seat.

The reason, found Katherine Whitcome and her colleagues at the University of Cincinnati, is down to the shape of the spine. The bones in women's lower backs are more wedge-shaped than men's, making it easier for the backbone to bend. The bony projections surrounding each little bone are also relatively larger in women, providing additional support to the spine. Because of this, women's spines have a more pronounced curvature than men's, which allows them to shift the upper half of the body backwards during pregnancy and counterbalance the weight of the growing baby.

14

Why do women get a linea nigra and other brown patches on their skin during pregnancy?

THE LINEA NIGRA is a brownish line extending from the belly button down to below the panty-line that appears in most women during pregnancy and remains for about a year after the birth. It corresponds to a line of collagen-rich connective tissue called the linea alba which separates the abdominal muscles and is what underlies the 'six-pack' you sometimes see on fit, muscular people.

One annoying side effect of rampaging pregnancy hormones such as oestrogen and progesterone is that they cause skin cells called melanocytes to churn out more of the brown pigment melanin (the same thing that makes you turn brown in the sun). These melanocytes are naturally more common in some areas of the body than others, including around previous areas of sun damage and along the lower half of the linea alba, which can leave many women with uneven patches of brown skin (also known as the mask of pregnancy) and this characteristic brown line running down their bellies.

Just why these cells should choose to hang out along the linea alba is unknown, but one theory is that many of them simply become stuck here during embryonic development, when they migrate out of a structure called the neural crest which runs down the middle of the embryo's body.

The mask of pregnancy usually disappears after pregnancy, but you can also prevent it from developing in the first place by wearing a good sunscreen containing both UVA and UVB protection.

Another irritation of pregnancy is that oestrogen combined with an increase in blood volume causes the growth of new blood vessels, and this can give rise to spider veins – small red branches on the surface of the skin. Most spider veins should fade after giving birth, but some may stubbornly remain. If they make you really unhappy, they can be removed with simple laser treatment, but it's probably worth waiting until your childbearing days are over, as it's likely you'll get more in subsequent pregnancies.

Added pressure from the unborn baby on the veins that return blood from the legs to the heart can also cause varicose veins, which afflict around 40 per cent of pregnant women. Elasticated stockings, sleeping on the left side of your body, light exercise, elevating your legs in the air and avoiding standing or sitting for long amounts of time can all reduce the chances of them developing.

15

Is there anything I can do to prevent stretch marks?

MY WAIST GREW 36cm (14in) during the course of my first pregnancy – not quite a doubling of its pre-pregnancy size, but if we assume the height of a pregnant bump is around 15cm (6in) that's still a sizeable increase. What's incredible is that you don't grow any new skin to do this; it's achieved simply through the stretching of existing skin.

Hormones including oestrogen and relaxin cause the collagen fibres that give skin its elasticity to become less sticky, so they can be pulled apart as the baby grows. Unfortunately, if this process happens too quickly, the result is stretch marks – a form of scarring. One recent study found that women with low levels of relaxin were more prone to stretch marks, possibly because their connective tissue was less elastic and therefore more prone to scarring. If you get through pregnancy without any stretch marks, you're in the lucky minority – particularly if you have white skin, which shows stretch marks more clearly.

Your susceptibility to stretch marks is largely dictated by your genetics and how quickly you grow, so you should try to keep weight gain as slow and steady as possible. Having strong abdominal muscles should also reduce the rate at which your bump pops out and therefore how quickly the skin stretches, adds Jenny Murase, a dermatologist at the University of California in San Francisco.

Many women try rubbing oil or cocoa butter into their bumps to ward off stretch marks, and several recent studies have set out to investigate whether this works. One, which analysed results in women who rubbed olive oil into their bumps twice a day for eight weeks during their second and third trimesters, found no significant difference in stretch marks, compared to women who did nothing. Two other studies which compared women who rubbed cocoa butter on their bumps with those who applied a standard moisturizer found no difference.

However, even if moisturizing doesn't prevent stretch marks, it may speed up their disappearance, says Murase. 'For my own pregnancy I used cocoa butter on my abdomen, even though there's no good published evidence that I'm aware of,' she says. 'Stretch marks are often reddened at first, and if the skin is overly dry it may be more irritated and take longer for that redness to fade, so I think it's a good practice to use moisturizer.'

There may yet be some hope in a cream called Trofolastin, which reduced the occurrence of stretch marks in women who applied it daily from week twelve of pregnancy, compared to those who used a placebo cream. According to the study, 56 per cent of women using the placebo developed stretch marks, against 34 per cent of those who used Trofolastin (which is marketed by Novartis and is available over the internet). It contains vitamin E, collagen, elastin, and extracts of a plant used in traditional Asian medicine called Gotu kola (Centella asiatica).

Another preliminary study suggests that women with higher levels of vitamin C in their bodies during late pregnancy are less likely to have stretch marks. However, since several studies have linked vitamin C supplements to preterm birth, it may be wiser to eat a healthy and balanced diet with plenty of fruit and vegetables than to cram yourself with vitamin C tablets.

16

Do big parents have bigger babies?

WHAT DO YOU get if you cross a male Shire horse with a female Shetland pony? If you're wincing at the very thought, the answer is a Shetland-sized foal that the mother can push out without being torn in two. Women with strapping partners needn't worry about how their brute of a child will fit through their narrow hips, because women have the greater say over the size of the baby they'll produce – at least up until the point of birth.

It was back in 1938 that researchers first thought of crossing

Shire horses with Shetland ponies in order to see what happened. They found that when a female Shire horse was mated with a male Shetland pony, the foal was bigger than a purely Shetland foal but smaller than a purely Shire foal. But when a female Shetland pony was mated with a male Shire horse, the foal weighed the same as a pure-bred Shetland foal. In other words, something was happening in the uterus of the female horse to control the foal's growth so it ended up in proportion to the mother's size. Of course, this is sensible from an evolutionary point of view; otherwise, many more women with large partners would have perished in labour, along with their babies.

More recently, the use of donor eggs and surrogacy to help infertile couples has shown that the newborn's size bears more relation to that of the biological mother, rather than that of the father or surrogate mother.

The current best guess of what's happening is that even though the foetus inherits a set of genes from both the mother and the father, the way they are expressed is not random, and, in some situations, the mother's genes take precedence and veto the father's genes. Generally, the mother's genes are thought to influence the size of the foetus, while the father's control the development of the placenta.

But Mum's diet can also influence how big her baby grows. On average, babies weigh an extra 26g for every kilogram a mother gains on top of the recommended weight gain of 25–35lb during pregnancy. High blood sugar caused by gestational diabetes can also lead to exceptionally large babies.

And if you had a big baby the first time around, watch out, because, on average, subsequent babies tend to be around 200g bigger. This may be because the uterus is already primed to grow new blood vessels, meaning more nutrients get through to the baby in subsequent pregnancies, helping them to grow bigger and stronger.

17

Why do some women show more than others? And why do women show earlier on in second pregnancies?

ALTHOUGH NO ONE seems to have studied this phenomenon, it's widely accepted that women on their second or third pregnancy tend to look a month 'bigger' than first-timers at the equivalent stage and usually have to dig out their maternity jeans much sooner.

Normally, the abdominal muscles on either side of the body are held together by a strip of connective tissue, but during pregnancy a hormone called relaxin weakens that tissue, and as the growing baby pushes against the abdominal muscles from the inside, they begin to separate.

Andrew Satin, professor of obstetrics at Johns Hopkins University in Baltimore, Maryland, says that in his general experience, women with stronger abdominal muscles tend to show less, and their abdominal muscles are less likely to be pulled apart during pregnancy. Women with smaller body frames are also likely to look bigger compared to tall, broader women.

The most plausible explanation as to why second-time mums show sooner is that their abdominal muscles are already weakened and stretched from the first pregnancy, and so provide less resistance against the hungry baby that is growing inside them, adds Stephanie Prendergast, a physiotherapist who specializes in postnatal care at the Pelvic Health and Rehabilitation Center in San Francisco. 'As a general rule, they also tend not to be in as good a shape as they were prior to the first baby, because they're already caring for a baby and don't have time to exercise as much,' she says.

18

Does a woman's body shape permanently change after pregnancy?

AS YOUR BUMP grows bigger, you may notice that it isn't just your waistband that seems to be getting tighter. During the first six months of pregnancy, most healthy women can expect to gain an extra 3.5kg of fat, which will mostly be deposited on the hips and thighs as an energy reserve for breastfeeding.

The good news is that, ultimately, your legs and bum may end up skinnier than they were before you got pregnant – although many women end up slightly heavier overall.

Unlike men, young women tend to pile 15–20kg of fat on to their hips and legs during puberty, which often stubbornly refuses to shift, regardless of how much exercise they do. But during the last ten to twelve weeks of pregnancy, and certainly once the baby is born, some of these fat reserves begin to be mobilized. Well-nourished breastfeeding women lose around 0.8kg of fat per month, much of it from previously protected reserves on the hips and thighs.

William Lassek and Steve Gaulin at the University of California at Santa Barbara analysed data from 16,635 American women and found that, although they tended to gain weight with each successive pregnancy, the women's hip and thigh circumferences decreased with every child they had. If they regained any weight after pregnancy, it tended to be deposited on their waists and upper bodies.

Lassek points out that fat mobilized from the hips and thighs provides a good source of an omega-3 fatty acid called DHA, which is critical for brain growth. 'During the first year of life, the brain increases by almost 1kg,' he says. 'Mobilized maternal fat may provide critical fatty acids and energy for brain growth.'

Although many women are overweight, Western diets are often deficient in DHA (which is found in cold-water fish such as salmon). A well-rounded backside is therefore good for your baby's health, and something to be embraced. And you can potentially look forward to a new wardrobe of cropped shorts and skimpy skirts, once the fat melts away.

19

Can jacuzzis and saunas really cause miscarriage?

YOUR GROWING BABY is insulated from the outside world by a thick layer of fat and muscle, so any changes in outside temperature are going to take a while to filter through.

However, sitting in a hot tub or sauna for a long time will cause your body temperature to rise, and there is some evidence that if it stays over 101°F (38°C) for a long time, particularly in the first trimester, this can increase the risk of miscarriage and birth defects.

It's important to note that most studies suggesting links between hot tubs and birth defects or miscarriage haven't studied the relationship between the two directly, but either looked at women who had prolonged bouts of fever during pregnancy (and these illnesses may also have influenced the outcome of the pregnancy) or conducted their research on animals. The most convincing human study found a rate of two to three neural-tube defects per thousand women among those who had been in a hot tub during the first trimester, compared to the normal rate of one defect per thousand women. However, this study didn't record how hot the water was, or how long the women stayed in for. Also, it asked women about their hot-tub use after their baby had already been diagnosed with a birth defect, so they may have been more likely to recall the experience than women with healthy babies.

If you do decide to use a hot tub when pregnant, the American College of Obstetricians and Gynecologists advises never letting your core body temperature go above 102.2°F (39°C). Since most hot tubs are around 104°F (40°C) and it only takes ten to twenty minutes of sitting in a hot tub to raise your body temperature to this level, women are often told to limit their time in a hot tub or sauna to just ten minutes.

Of course, your body temperature will begin to drop as soon as you step outside the tub, and even non-pregnant women tend to exit the sauna or hot tub when they begin to feel uncomfortably hot – a good sign that your body temperature is rising.

Taking a warm bath during pregnancy is a different matter – the top half of your body is not usually immersed and the water is not constantly being heated, so it takes longer for your body to heat up.

Rumours also circulate that bathing during pregnancy

might increase your chances of an infection that could trigger miscarriage, by bacteria in the water making their way up the vagina and into the uterus. During pregnancy, however, a thick plug of mucus in the cervix effectively seals off the uterus from the outside world, making this extremely unlikely. Some studies have hinted that this may be more of a problem once a woman's waters have broken, although the evidence is weak (see **49**: 'Can I take a bath once my waters have broken?').

20

Is it safe to have sex?

NORMALLY, THE VAGINA is linked to the uterus by a narrow tunnel through the cervix, which the sperm must navigate in order for a woman to become pregnant. However, sperm have a hard time getting through once a woman is pregnant, because a thick plug of mucus blocks the tunnel. This plug also seals off the uterus from the millions of bacteria living in the vagina, reducing the risk of any infection that could damage a growing baby. One of the first signs of labour is the release of this plug – known as the 'show' – when a blob of thick, occasionally blood-stained mucus appears in your knickers.

Even if any semen did get through the mucus plug, the baby is encased in a tough pair of membranes, called the amniotic sac, so it's extremely unlikely that any semen could ever reach the baby.

21

Why do so many human embryos miscarry before twelve weeks?

IT'S A STATISTIC THAT haunts most women during the early weeks of pregnancy: as many as one in three pregnancies end in miscarriage in the first trimester. I spent the first months of both my pregnancies obsessively checking off the days until my twelve-week scan, convinced that when I got there the sonographer would find a blank space where my baby was supposed to be. I also have friends who have been through the trauma of successive miscarriages, and the overriding question they ask is whether there is anything they could have done differently. According to the latest thinking on the main causes of miscarriage, the answer is almost certainly no.

'I don't believe that jogging around the block, or having sex, or having a glass of wine, or eating unpasteurized cheese is going to knock off an embryo that's destined to be successful,' says Lesley Regan, an expert in recurrent miscarriage at Imperial College, London. 'It's usually just coincidental.'

A major cause of human miscarriage is that an egg that contains an abnormal number of chromosomes has been fertilized, also known as aneuploidy. Such embryos may grow for a while, but they often stop growing and abort during early pregnancy (often before a woman even knows she is pregnant). Humans seem particularly prone to this problem compared with other mammals, though no one really understands why. For a woman between the ages of thirty and thirty-five, about 60 per cent of human eggs are aneuploid, and this rises to more than 90 per cent for a woman in her early forties. 'Over the age of thirty-five, the miscarriage rate goes up dramatically and the fertility rate goes down, so many, many pregnancies are lost,' says Regan.

As well as helping to explain why so many pregnancies fail, this helps to explain why it often takes so long to get pregnant in the first place – particularly as we age. Even among young healthy couples who diligently have sex several times around the day of ovulation, it can take months for the woman to fall pregnant. It's not necessarily that she isn't conceiving but that huge numbers of embryos are lost before a pregnancy test would even detect their existence.

Some of the commonly cited statistics about miscarriage before twelve weeks can be misleading because they are heavily dependent on when you take a pregnancy test. Take the 'one in three' figure. This seems to come from a study which confirmed pregnancy at the earliest point possible (before most commercial pregnancy tests would show a positive result). It found that 22 per cent of pregnancies ended before they would have been detected using a normal pregnancy test, and 31 per cent of pregnancies were lost overall. In other words, it isn't relevant to the everyday woman who has just found out that she's pregnant after doing a test she bought at the pharmacy, because if the pregnancy has progressed far enough for it to be detected using a home-test kit, the risk of miscarriage has already dropped significantly.

Slightly more accurate is the 'one in five' figure provided by the UK's Royal College of Obstetricians and Gynaecologists. This refers to women who have done a pregnancy test and got a positive result. Although the rate may still feel worryingly high for most couples, it's also worth bearing in mind that it refers to the general population rather than individuals, and some women are at greater risk of having a miscarriage than others. Age is a factor, as is having had a previous miscarriage, while certain medical conditions such as diabetes or thyroid disease can also increase your risk.

More than twenty years ago, Regan conducted a study to assess the risk of miscarriage among different groups of women. Once pregnancy was suspected, the women came in for an ultrasound scan, and these scans were then repeated every two weeks until they reached week twelve. Overall, she found that 12 per cent of the women miscarried but, for first-time pregnancies, and for women whose last pregnancy resulted in a healthy baby, the risk of miscarriage was more like 5 per cent – while for women who had carried several healthy babies to term, the risk dropped to 4 per cent.

Others were at greater risk of miscarriage. Among women whose last pregnancy had miscarried, the risk was 19 per cent, while for women who had experienced several miscarriages their risk of miscarrying in this pregnancy was 25 per cent (although this still, of course, meant that 75 per cent of them had healthy babies).

The message from all of this is pretty clear. If this is your first baby, or your previous pregnancies have been healthy and you are showing no symptoms of miscarriage (such as pain or bleeding), then there is a far higher chance that you will carry a healthy baby to term than suffer the tragedy of miscarriage. That risk also drops the later in pregnancy you progress.

Exercise

22

Is it safe to exercise during pregnancy?

MANY WOMEN STOP doing vigorous exercise such as jogging during pregnancy for fear that it will increase their risk of miscarriage or damage their baby in some other way.

Although at least one large study has suggested that high-intensity exercise for more than seven hours a week increases the rate of miscarriage, most of the data was collected after miscarriages had occurred, so women may have recalled their exercise levels differently, or been more motivated to take part in the study in the first place. Other studies have hinted that strenuous exercise might temporarily divert blood away from the foetus, meaning it gets less oxygen.

On the other hand, there is also some preliminary evidence that exercise during pregnancy results in leaner babies with advanced brain development and lower heart rates – something that could have lifelong benefits for health.

Current guidelines in the USA, UK and other countries advise women that they can continue exercising at a similar level of intensity to before they got pregnant, unless they have certain medical conditions, such as heart disease, or they have been told their pregnancy is at risk, because of prolonged bleeding, for example. When the Cochrane Collaboration explored the issue in 2009, it concluded that there is insufficient evidence to say whether exercise is good or bad for the baby, although aerobic exercise two to three times a week would improve or maintain the mother's fitness level.

However, emerging evidence suggests that even strenuous exercise during pregnancy isn't harmful, provided that the exercise programme is tailored to the mother's level of fitness. Andrew Satin at Johns Hopkins School of Medicine asked heavily pregnant women to walk or run on a treadmill until they reached their maximal exertion threshold – the point where they can go no further. All the while, the movement and heart rate of the baby was monitored for any problems or signs of distress, as was blood flow to the uterus and amniotic fluid levels.

Even in women who didn't exercise before pregnancy, no impact on the baby's health was observed. 'Contrary to what we believed in the past, it is even OK for women who have previously been inactive to initiate exercise during pregnancy,' says Satin.

Although you might imagine that jogging would cause a baby to be flung around in all directions, Satin was surprised to see little change in the baby's position. 'The uterus moves up and down and the baby moves with it, but we really didn't observe radical changes of position at all,' he says.

Above all, women who choose to exercise while pregnant should listen to their own bodies, and slow down or stop if they feel light-headed or unwell. Pregnancy is not a good time to push through the pain and keep going. Satin also advises drinking plenty of water to prevent dehydration, and to avoid any activity where you might fall or get hit, such as skiing, horse-riding, mountain biking or martial arts. This is particularly important as the baby grows bigger and ceases to be protected by the bones of the pelvis (typically, after around twelve weeks of pregnancy), when there is a risk that a serious knock could cause the placenta to become detached from the uterine wall, putting both mother and baby at risk.

23

What stomach exercises can I do when pregnant?

ALTHOUGH YOU SHOULD probably avoid lying on your back for prolonged periods during the latter half of pregnancy, this doesn't mean you can't exercise your abdominal muscles. In fact, it's important to maintain strong abdominal muscles during pregnancy because these help support the pelvic floor and maintain your posture, which can be thrown off balance by that ever-bulging bump.

Exercises that should be avoided are full sit-ups, abdominal crunches, and those that involve twisting over to one side. This is because the central abdominal muscles begin to pull apart during late pregnancy, and these exercises can widen that gap even further. (For tips on how to close the gap once your baby is born, see 71: 'Can other countries (e.g. France) teach us anything about getting back into shape after birth?')

However, one exercise that is OK during mid to late pregnancy is the transversus exercise, which can be done in any position and involves placing both hands below your belly button, taking a deep breath in and then pulling the abdomen and belly button in towards the spine on the out breath. Hold this contraction for up to ten seconds, breathing normally while you do it, and repeat up to ten times.

Exercises such as the plank are also fine. Lying on your front, you pull your belly button in tight and push up off the floor with your hands or elbows shoulder-width apart and your leg muscles engaged. You can also do a side plank, where your weight is supported on one hand or elbow and your feet. Both of these exercises can be made easier by bending your knees and using these to support your weight rather than your feet.

Pelvic tilts can help to strengthen the back as well as the abdominal muscles. Lie on your back with your knees bent and your feet resting on the floor, then tighten and pull in the abdominal and buttock muscles while pressing the small of your back into the floor.

24

Will lying on my back harm my baby?

DURING LATE PREGNANCY, the added weight of the baby, amniotic fluid and the placenta can start to compress the large vein that carries blood from the lower half of the body back to the heart, pooling blood in the legs. Rather than harming the baby, the most likely impact of this is that it will make the mother feel light-headed or dizzy, so she will want to turn over – although if she continues to lie on her back then this might ultimately reduce the amount of blood reaching the baby. However, I could find no published evidence to prove that lying on your back harms your baby.

Even in non-pregnant women, there is a slight decrease in the amount of blood the heart pumps when you lie on your back, compared with when you lie on your left side. One recent study that compared heart function in pregnant and non-pregnant women found a further drop once women reached around twenty weeks in their pregnancy, so this seems a sensible time to start to avoid lying flat on your back for prolonged periods.

Because this vein is on the right side of the body, lying on your right side can have a similar effect to being on your back. This may be why many women say they feel more comfortable lying on their left side as their pregnancy progresses.

'I think the key word here is "prolonged",' says Linda Szymanski, a professor of obstetrics at Johns Hopkins School of Medicine. A couple of minutes on your back or right side is unlikely to cause any harm, but if you feel dizzy or light-headed you should roll on to your left side.

25

Do pelvic-floor exercises actually work?

HERE'S A REASON to start clenching those pelvic-floor muscles: they can boost your sex life. Your pelvic-floor muscles wrap around the inside of your pelvis, supporting and helping to control your bladder, vagina and rectum. Midwives are constantly nagging pregnant women to do their pelvic-floor exercises – I still have red stickers dotted around my house to remind me to squeeze them while watching television, boiling the kettle or brushing my teeth. The reason usually given for doing them is that these exercises can stave off the risk of developing urinary and fecal incontinence, which sadly affect a large number of women (at least temporarily) after birth. Up to 38 per cent of women have some degree of urinary incontinence in the two to three months after birth – often a small amount of leakage when they laugh or sneeze – and up to 6 per cent find it difficult to control their bowels. A recent review concluded that women who do pelvic-floor exercises during pregnancy are 56 per cent less likely to develop urinary incontinence during late pregnancy and in the six months after birth than those who don't.

What fewer people know is that if you carry on doing these exercises after childbirth, it could improve your sex life. In a recent study conducted in Turkey, a group of seventy-five women who had given birth four months previously were divided into two groups, and half of them were taught how to exercise their pelvic-floor muscles. They started off contracting them for three seconds followed by a three-second rest, then for two seconds ten times a day for fifteen days, after which they progressed to holding the longer contraction for five seconds, and eventually ten, and increasing the number of workouts to fifteen times a day.

By month seven, women who had been exercising their pelvic-floor muscles found it easier to become sexually aroused, were better lubricated, had more intense orgasms and were more sexually satisfied than women who hadn't done the exercises. The pelvic-floor muscles are responsible for the rhythmical contractions of female orgasm, but they can also influence how well women respond to sexual contact, by affecting the location of the inner section of the clitoris, for example (see 71: 'Can other countries (e.g. France) teach us anything about getting back into shape after birth?').

Sadly, the evidence suggests that pelvic-floor exercises are unlikely to help you have an easier birth or reduce the risk of tearing. But, given the other benefits, it seems worth the effort.

26

How do I know if I'm doing my pelvic-floor exercises properly?

SOMETIMES WOMEN THINK that they are clenching their pelvic-floor muscles, when in fact they're clenching other muscles such as those of the buttocks. One way to check is to lie naked on your back in front of a mirror so that you can see the entrance to your vagina and anus. First, breathe in and relax your muscles, then start to squeeze from the back passage and up towards those muscles you would be using if you were trying to stop the flow of urine. You should see the anus contract, the area of tissue between the vagina and anus start to lift and the entrance to the vagina start to close. 'It should be a nice slow upward and inward movement,' says Maria Elliott of SimplyWomensHealth in London, who specializes in pre- and postnatal physiotherapy. If you don't see this, it suggests that you're not tightening the correct muscles, so you should either try again or speak to a health professional.

Another tip is to use your breath to strengthen the force of pelvic-floor-muscle contractions. 'You will get a much better pelvic-floor contraction on the out breath,' says Elliott. 'Long term, the pelvic-floor muscles work with the abdominal muscles and diaphragm like a muscular cylinder. Doing squeezes down there is OK, but ultimately the pelvic floor needs to start working as a team with those other muscles to get coordinated activation.'

Most hospitals recommend doing five sets of longer contractions and five to ten sets of short contractions five times a day, but the precise pattern varies. Instead, Elliott suggests cutting back on some of these and replacing them with some long, slow, breath-controlled contractions instead. To do this, take a deep breath into your ribcage and, as you breathe out, first pull in the pelvic-floor muscles and then zip in the belly button as if you are trying to squeeze into a tight pair of jeans. Relax as you breathe in, and then repeat as you breathe out again. Elliott recommends doing twelve of these out-breath contractions three times a day, ideally while lying on your side in bed or on the sofa, as soon as possible after having had your baby. Avoiding constipation and heavy lifting during the first eight weeks is also recommended to avoid prolapse – a nasty condition where the uterus begins to fall into the vagina.

Even if you don't remember to exercise your pelvic-floor muscles every day, the important thing is that you're doing some exercises rather than none. So get squeezing!

Baby on the Brain

27

Does pregnancy make women forgetful?

LOSING TRACK OF what you were saying halfway through a sentence? Walking upstairs only to forget what you went there for in the first place? Many women claim to become more forgetful during pregnancy, but evidence for the existence of 'mumnesia', or pregnancy-related forgetfulness, is mixed. Some studies have shown that pregnant women score worse on tasks involving memory storage and retrieval, while others have found no correlation between pregnancy and forgetfulness.

One possibility is that women simply report being more forgetful because we expect pregnancy to scramble our brains so are more aware of any lapses in memory. However, brain scans have also shown that women's brains shrink during the final months of pregnancy and don't return to their previous size until around six months after birth – possibly because of the effects of hormones, or because the foetus is poaching nutrients for itself. This at least provides some evidence of tangible changes to the brain. On the other hand, we don't know if this shrinking actually has any effect on brainpower, or if it affects all areas of the brain equally. In fact, experiments in rats have shown that the brain's memory centre, the hippocampus, grows new neurons during pregnancy, so even if there is a short-term lapse in memory, there may be other, long-term benefits.

Even in the studies that have found evidence for mumnesia, not all types of memory are affected equally. In one of the largest studies to date, 254 pregnant women were asked to do a series of memory tests, which were then repeated twelve to fourteen weeks after the birth. The tests were then carried out on forty-eight non-pregnant women. Women in late pregnancy and new mums fared worse than non-pregnant women when it came to memorizing pairs of words (verbal memory). But working memory (or the ability to manipulate information already stored in the brain) and the ability to recognize faces were unaffected. Also, not all women were affected equally.

It's also possible that what pregnant women lose in memory they make up for in other areas. 'The changes that are occurring may make you a better mother,' says Laura Glynn at the University of California at Irvine, who led the study. Oestrogen, in particular, aids the development of maternal instinct and nurturing, and high levels during pregnancy have been linked to increased feelings of attachment to children after birth.

In rats, at least, pregnancy and motherhood also seem to trigger improvements in navigation skills and memory for places, which might make female rats better at foraging food. Meanwhile, researchers at the University of Bristol, UK, have found that women's sensitivity to facial expressions showing fear, anger and disgust increase during pregnancy, perhaps training them to become hyper-vigilant to threats once their children are born.

As pregnancy wears on, women also seem to become less

fazed by stressful situations. A classic way of testing stress in the laboratory is the Trier Social Stress Test – essentially, a mock job interview, during which your body language and levels of key hormones are monitored. A study of 150 pregnant women found that the more advanced their pregnancies, the less stressful they found the test to be. Women in late pregnancy also had lower blood pressure, slower heart rates and lower increases in cortisol during the test than women at twenty-one weeks of pregnancy and non-pregnant women.

We still don't know whether any of these changes are permanent, and what other beneficial changes might occur to the brain during pregnancy. 'It is possible that when faced with different tasks related to the care of offspring, such as multi-tasking, performance under stress or sensitivity to infant cues, mothers' performance might be enhanced,' says Glynn.

28

Is stress during pregnancy bad for my baby?

IF YOU BELIEVE what you read in the newspapers, stress during pregnancy trebles your risk of miscarriage, doubles your risk of stillbirth, gives your child schizophrenia, asthma and attention deficit hyperactivity disorder (ADHD), inhibits their emotional growth and lowers their IQ. Indeed, if you type 'stress' and 'pregnancy' into Google, you get close to 77 million results. It's enough to raise your blood pressure just thinking about it.

But before you pack in your job and lie in bed for the rest of your pregnancy, you should know that many of these studies relied either on animal studies or on women recalling how stressed they were rather than measuring their stress directly; or they involved events far more stressful than taking a crowded bus to work or pulling an all-nighter to meet a deadline. The timing of the stressor and how you react to it also seem to be important factors.

There are known mechanisms by which psychological stress could damage an unborn child. One possibility is that women who are stressed are more likely to smoke, drink alcohol, eat badly or not make it to their antenatal check-ups. But stress also prompts the release of several hormones, including cortisol, which helps your body cope with stress by providing a quick energy boost and heightened awareness. Usually, cortisol production is quickly shut off, but in very stressful circumstances such as losing a relative or living through an event such as a war or an earthquake, this doesn't happen, leading to chronically high levels. Studies in animals have shown that high cortisol levels can shrink areas of their baby's brain involved in memory storage, as well as trigger miscarriage and preterm birth. Cortisol and related hormones can also decrease blood flow to the uterus, and dampen the immune system, leaving you more prone to infection.

However, stressing animals in a lab isn't the same as understanding how stress affects pregnant women. Human studies have produced far more mixed results, with some suggesting there is an effect and others showing no effect. One of the biggest problems is defining what stress is, but there are also issues of timing. There is some evidence that women who lived through a big earthquake during pregnancy may have given birth prematurely more frequently than other women – but only in those women who experienced the earthquake during their first trimester. These women also described the event as more stressful than women in their second and third trimesters. Similarly, pregnant women living near the World Trade Center in New York on 11 September 2001 had slightly more preterm births than women living elsewhere, but only if they were in the first trimester at the time of the attacks. However, among women who experienced the Chernobyl disaster in Ukraine, there was no increased risk of preterm labour.

One way to get around this problem is to do something called a meta-analysis, which involves combining the results of numerous independent studies to see what they say on average. When Heather Littleton at East Carolina University recently did this with thirty-five studies on stress and pregnancy, she found that babies born to stressed mothers were slightly more likely to have a low birth-weight or to be underweight in the weeks after birth, but the effect was tiny. 'On average, these studies are finding that stress only accounts for 1 per cent or less of the variability in individuals' outcomes, such as the weight of their babies,' says Littleton. The other 99 per cent is down to other things, such as genetics and social or lifestyle factors.

Another way of investigating how stress affects unborn babies is to study it directly. Janet DiPietro and her colleagues at Johns Hopkins University examined 112 healthy pregnant women living in the US three times during their third trimester. They asked the women about their stress levels and recorded how much the baby was moving. They also examined their babies two weeks after birth.

The babies of stressed mums seemed to be more active and, after birth, they were more irritable, but they also scored higher on a brain-maturation test and had better control of their body movements.

Since cortisol is known to play a role in normal brain maturation, it's possible that moderate, short-lived stress during pregnancy might boost this process and even be good for babies. DiPietro says that pregnant women should take it easy but not worry about moderate levels of everyday stress. 'Relaxation is good during pregnancy – if you like to relax. But there is no evidence that everyday stress has any adverse effects on the baby's development,' she says.

29

Can men get pregnancy symptoms too?

NAUSEA, HEARTBURN, ABDOMINAL PAIN . . . If you thought it was just women who suffer these symptoms during pregnancy, think again. Although it's not officially recognized as a disease or mental illness, somewhere between 11 and 50 per cent of men are thought to experience symptoms of couvade syndrome, or sympathetic pregnancy. The most common symptoms include loss of appetite, toothache, nausea, sickness and anxiety. Symptoms tend to peak in the third month of pregnancy, when many women are suffering the evils of morning sickness, then diminish in the second trimester, only to rise again during the final month.

The term 'couvade' derives from the French verb *couver*, meaning to hatch or sit on eggs. People have put forward various theories for why couvade syndrome might exist, including the man seeing the unborn child as a rival for his partner's affection and therefore unconsciously trying to attract her attention, or it being a physical manifestation of the identity crisis triggered by the prospect of becoming a father.

Although very little research has been done on the causes of couvade syndrome, those studies that do exist suggest that men are more likely to suffer symptoms if they are highly involved in their partner's pregnancy and in preparing for the birth.

If couvade syndrome sounds like just another form of male hypochondria to you, there may yet be a physical source of the symptoms. One study found higher levels of the hormone prolactin and lower levels of testosterone in men reporting more than two couvade symptoms such as fatigue, altered appetite and weight gain compared to men with no symptoms – although just how these hormonal changes might trigger pregnancy-like symptoms in men remains unclear.

However, regardless of whether you buy into the existence of couvade syndrome, there is a pregnancy-related illness that men do suffer from and which should be taken seriously. Postnatal depression is estimated to affect between 4 and 25 per cent of men in the two months following the birth of a child, although its onset may be more gradual than in women.

Having a partner who is experiencing postnatal depression is the greatest risk factor in men developing it, and just as maternal postnatal depression can impact on the baby's emotional development, paternal depression has a similar effect.

30

Why do women go into nesting overdrive in the final weeks of pregnancy?

I SPENT THE final days of my first pregnancy baking, painting the nursery walls, sorting and re-sorting my daughter's new clothes, not to mention sewing my first (and only) patchwork blanket to go in her cot. In my second pregnancy I dismantled a second-hand double-buggy and spent hours obsessively scrubbing the fabric parts to get them clean. It is often said that women go into nesting overdrive in the final weeks of pregnancy, and hormonal changes may help to explain this. For most of a pregnancy, progesterone dominates over oestrogen, but as the birth approaches, oestrogen levels begin to rise, and studies have found that women experiencing a greater shift in oestrogen report more feelings of attachment and well-being once their babies are born. High levels of the hormones oxytocin and prolactin also prompt the onset of maternal behaviours, such as becoming hyper-responsive to anything baby-related, raising the pitch of your voice and becoming generally more touchy-feely.

Studies of rats hint at how these hormones might affect the brain on a physical level. As birth approaches, several structures at the front of the brain which have been linked to mothering seem to grow in size. These include brain areas associated with addiction and reward, the processing of emotions, and empathy, or the ability to read the minds of others. Pregnancy also triggers the growth of new brain cells in areas associated with learning and memory, and, in rats at least, these changes seem to be permanent: boosting their memory for places, their foraging abilities and making them less anxious in the face of challenges. This pattern of new cell growth is reinforced with each subsequent pregnancy. Assuming the same changes take place in humans, this means that the more kids you have, the more 'mummy-fied' your brain will become as you shift from a life that largely centred around your own needs to one where others now take precedence. Most mums agree that they never stop worrying about their child or children, no matter how old they get. I also feel as if I'm more vigilant and aware of threats now I'm a parent, and I become claustrophobic if I take my children into a busy shop.

The hormone we know most about is oxytocin – often referred to as the 'cuddle chemical' but more accurately described as a hormone involved in forming relationships with others. Levels of oxytocin are boosted in pregnant women and remain high once the baby is born – and possibly for much longer (although no one has measured it beyond three years). Those same brain areas associated with emotional processing and reward are highly responsive to oxytocin, and elevated levels of it are related to more sensitive parenting in both mums and dads.

Although these hormonal shifts may prepare women to feel maternal once their child is born, the system is also reinforced by experience. Oxytocin is released in response to cuddling and spending time with your baby, so the more you play with your child, the more your natural parenting instinct will be activated.

Breastfeeding also triggers the release of oxytocin. A recent study that compared oxytocin levels in breastfeeding mothers with women who were bottle-feeding their infants found higher levels in the breastfeeding women. When the same women were played a recording of either their own or a different infant crying, the researchers also noticed greater activation in the brains of breastfeeding mums in response to their baby's cry. Although this doesn't necessarily mean that breastfeeding makes you a more sensitive mother, it does show how important oxytocin is in driving maternal behaviour.

31

Are some women naturally more maternal than others?

IF YOU SCAN the brain of a woman three weeks after she has given birth, you will see greater activation of the areas involved in emotional processing, reward and addiction – precisely the same areas that are stimulated when you fall in love. As different as maternal and romantic love may seem, on a physical level having a baby is almost identical to meeting the partner of your dreams.

Obsession is one example. An area of the brain, the striatum, which is linked to obsessive behaviour, is activated both when we fall in love and when we have a baby. Anyone who has been in love will remember those heady early days when you're constantly replaying conversations in your mind and checking to see if your lover has called. Mums and dads also show symptoms of obsessive-compulsive behaviour in the early days after their child's birth, such as ritually checking whether their baby is breathing or is warm enough.

However, not all women react to the birth of their baby in exactly the same way. Ruth Feldman at Bar-Ilan University in Ramat-Gan, Israel, has found that new mothers can broadly be divided into two categories based on their brain activity. She focused on two brain networks, one which links an area associated with emotional processing to social areas associated with empathy and reading cues given by babies and the other linking an area associated with reward and addiction to the same social areas.

Although mothers showed activity in both brain networks when interacting with their babies, some showed more activity in the reward and addiction network while others showed more activity in the emotional-processing network. The first group of women essentially became addicted to their babies, but they were also extremely in tune with their needs. Feldman calls them 'synchronous mothers'. 'Their main feeling is reward, energy and enjoyment,' she says.

For women showing greater activity in the emotional-processing areas, 'the overwhelming feeling is anxiety, looking for harm, and a sense of "I need to protect my infant",' Feldman says. This style of mothering could have evolutionary advantages – preparing a child to be vigilant of their surroundings and less trusting, for example. But the payoff is being less aware of the non-verbal signals their infants may be giving off, such as not noticing when their baby is tired. Feldman calls these women 'intrusive mothers'.

Although less common, there was also a third category of women, who showed little activity in either network in response to their babies. These women were suffering from varying degrees of postnatal depression.

It's also interesting that the children of these mothers showed differences in their behaviour at the age of three. 'They behaved towards their best friends in the same way as their parents behaved towards them when they were infants,' says Feldman. The children of synchronous mothers were more engaged with their friend's needs and better at sharing, while the children of intrusive mums were more fidgety and less able to tolerate tension. Meanwhile, children whose mothers had suffered from postnatal depression were more withdrawn and their social skills less developed.

As well as demonstrating the importance of recognizing and treating postnatal depression, this study suggests that women shouldn't necessarily expect to react in the same way to becoming a mum. Feldman also emphasizes that these instinctive behaviours can be changed with practice. For example, intrusive mums can learn to be more sensitive to signals such as their baby turning away from them when they'd like a rest. 'When you see that, just wait patiently for the child to look at you again,' Feldman says.

32

Do men change when they become dads?

MEN OFTEN GET a bad press when it comes to fatherhood. For centuries, they have been groomed to believe that their biological role is to provide for the family rather than to play any active part in their child's upbringing. However, several recent studies have challenged this, suggesting that men also experience biological changes when they become fathers and that these can make them happier, more engaged dads – if they allow themselves to be.

As well as researching how mothers' brains change in response to childbirth, Feldman has measured levels of the hormone prolactin in men during the later stages of their partner's pregnancy. This hormone is best known for its role in triggering breastfeeding, but Feldman has found that male prolactin levels also increase as birth approaches. What's more, men with higher levels of prolactin seem to be more alert and responsive to their infant's cries once the child is born.

In a separate study, Feldman measured prolactin levels in forty-three first-time dads when their babies were six months old. She found that men with the highest levels of this hormone were more likely to engage with their child, encouraging them to explore and interact with new toys. Men's prolactin levels also shoot up when they hear their baby cry.

But prolactin isn't the only hormonal change affecting new dads. Levels of testosterone also drop, which may make fathers less aggressive and more attentive to their family's needs. At least one study has found that men with higher testosterone levels are less likely to give their children attention, while another study found that fathers with lower testosterone expressed more sympathy and a greater need to respond when they heard their baby cry. Falling testosterone may also discourage men from looking elsewhere for sex while their wives are too busy breastfeeding to satisfy their needs.

Expectant fathers see increased levels of the stress hormone cortisol in the three weeks before birth, while new dads often experience a surge of the female sex hormone oestrogen when their babies are born. Animal studies suggest that this triggers nurturing behaviour in males.

Finally, Feldman has found that levels of the 'cuddle chemical' oxytocin increase in new dads, and this too seems to relate to paternal behaviour. Dads with higher levels of oxytocin played with their children more often and seemed more attached to them than those with less of the hormone.

Unlike women, for whom oxytocin levels automatically rise throughout pregnancy and as a result of breastfeeding, oxytocin production in men seems to be more hands on. In a separate study, Feldman found that fathers' oxytocin levels were boosted in response to playing and interacting with their children. In other words, when it comes to parenting, men reap what they sow. Probably, the more fathers engage with their kids, the more their paternal instinct will kick in, producing happier, more nurturing dads as a result.

The Developing Baby

33

When does a baby become conscious?

THROUGHOUT MY OWN pregnancies I've pondered the question of consciousness. Can a foetus make sense of its environment, and does it have a sense of identity? Can it think, dream or feel pain? And does the emergence of these senses happen suddenly, like turning on a light switch, or is it more of a gradual process?

These are difficult questions to answer, not least because we've yet to come up with a proper definition of what consciousness is in adults, let alone in unborn babies who have no way of describing their experiences to us.

A key moment is when the peripheral nervous system (all the nerves relaying movement and sensations from limbs and tissues outside the brain) joins up with the cerebral cortex (the brain region responsible for higher thought processes such as memory, awareness, attention and language). This connection, which occurs at around twenty-five weeks, is a vital link between the outside world and the higher brain. Without it, it is impossible to be aware of your surroundings or to make sense of them, says Hugo

Langercrantz, an expert on foetal consciousness at the Karolinska Institute in Stockholm, Sweden. This includes becoming conscious of pain – although it is clear that foetuses react to and move away from painful stimuli at much earlier stages of development.

So what has been running through the baby's brain before this point? By the end of the first trimester, when your baby no longer looks like a fleshy tadpole but is starting to look vaguely human, their brain circuitry is roughly equivalent to that of an earthworm or a marine snail, says David Edelman, a consciousness researcher at the Neurosciences Institute in La Jolla, California. Obviously, human brains are programmed to develop far beyond this, but twelve weeks after conception, the foetus has only got as far as building the circuits that allow it to respond to its environment by reflex, without any higher involvement from the brain.

By around seventeen or eighteen weeks, the nerves from the rest of the body begin to connect to the brain, bringing the baby's state of consciousness roughly in line with that of a reptile such as a snake. A reptile undoubtedly responds to its external environment, tracking down a mouse by sight and smell, for example, but most scientists don't think reptiles engage in higher thought processes such as abstract reasoning, because they don't possess a cerebral cortex (the outermost, folded layers of the brain). This is a moot point because neither do some birds such as ravens and parrots possess a cerebral cortex, yet they appear to possess some degree of language and an awareness of others.

At nineteen weeks, human embryos will withdraw from painful stimuli such as being poked with a needle. By twenty-three weeks, they release stress hormones such as cortisol, noradrenaline and a natural painkiller called beta-endorphin in response to a needle prick – even though higher brain regions aren't yet involved in processing these responses. If unborn babies do feel pain at this stage, it's unlikely to be the type of pain we experience as adults, but it may be pain nonetheless. It's a bit like trying to put ourselves in the shoes of an injured animal – it's impossible, because our brains are wired differently.

At twenty-five weeks, that all-important connection between the cerebral cortex and the rest of the body is made, making human-like reasoning possible. However, even now it would be foolish to try to ascribe adult-like consciousness to a foetus. None of us can remember what it was like in the womb, and one reason for this may be that we didn't have the words or experience to make sense of the various sensations feeding into our brains. Many experts now believe that consciousness isn't an all-or-nothing phenomenon but a gradual turning-up of the dimmer switch which continues throughout early childhood.

In addition, unborn babies are kept relatively sedated in the womb through a combination of low oxygen levels and anaesthetic-like chemicals produced by the placenta. Foetuses show cycles of sleep and wakefulness, and there is some evidence that they can learn, but only at birth do they fully wake up for the first time.

34

Does a wriggly bump equal a boisterous baby?

JUST SEVEN WEEKS into pregnancy, the developing baby makes its first movements. These start slowly, with gentle back bends, but by sixteen weeks the foetus will be capable of virtually the full range of movements that a newborn baby is capable of – in fact, more so, because it is not constrained by gravity. A sixteen-week-old foetus can hiccup, make breathing movements, yawn, touch its hand to its face, kick its legs, suck, swallow and move its eyes.

Despite all this, expectant mothers rarely start to feel any movements before eighteen weeks of pregnancy, and often much later. Some women describe it as a lurching motion, like being on a rollercoaster, while others say it feels like a gentle popping. Rest assured, your baby's movements will only get stronger as time wears on, and some babies are particularly keen on making their presence felt.

During the second half of pregnancy, the baby changes position approximately once a minute and is active around 30 per cent of the time. However, women perceive only around a tenth of their baby's movements, because the impact is buffered by the sac of amniotic fluid in which the baby floats. At twenty weeks, the baby takes up only about half of the available space in the uterus, so it has considerable room in which to kick, punch and spin around. By the end of the pregnancy, the baby takes up around 90 per cent of the available space – even though the uterus has stretched significantly – so it is considerably more restricted in its movements.

When I was about twenty-three weeks pregnant with my first child, Matilda, I took part in a study which involved lying in an MRI scanner while a group of researchers from Imperial College, London, took live video recordings of her movements. When they played them back to me, I was astounded to see her kung-fu-kicking the inside of my belly, crouching and jumping, when all I had felt was the odd pop. She was even making pincer movements with her thumb and forefinger: something she didn't do again until she was around seven months of age. When she was slightly older, I was constantly pummelled by her kicks, particularly in the early evening, after I had eaten. Now a spirited toddler, Matilda is one of the most active (and exhausting) children I know.

So it's intriguing to hear of several studies suggesting a correlation between a baby's activity in the womb and their temperament once they are born. Back in 2002, Janet DiPietro and her colleagues at Johns Hopkins University monitored the heart rate and body movements of fifty-two foetuses at twenty-four, thirty and thirty-six weeks of pregnancy, then examined the babies at one and two years of age.

They found that babies who were more active in the womb seemed to be less distressed when they were frustrated or restrained by their parents when aged one, and were more interested in interacting with toys and strangers when aged two. There was also a difference between boys and girls. Boys that were particularly energetic at thirty-six weeks of pregnancy went on to become more active children, while the opposite was true for girls.

In a separate study, Ian St James-Roberts and Praveen Menon-Johansson at the Institute of Education in London asked twenty women who were thirty-seven weeks pregnant to log their baby's movements for an hour every morning and evening for three days. After birth, the baby's sleeping, feeding and crying patterns were also recorded at one, six and twelve weeks of age.

Although the unborn babies tended to produce their strongest kicks in the evening, these movements didn't really match up with the baby's activity levels once they were born. What did match up were the more subtle movements that mothers could detect, particularly those occurring in the morning. Babies that moved a lot were more prone to crying, while those that were less active tended to be more peaceful.

As a general rule, the later in pregnancy you look, the more an unborn baby's temperament seems to match up with its older self.

But even very early in pregnancy there may be a hint of what's to come. Shimon Degani and his colleagues at the Bnai-Zion Medical Centre in Haifa, Israel, used ultrasound to study the activity of twenty-two pairs of twins between eleven and fourteen weeks of pregnancy and found that the more active twin also tended to be the more difficult, unpredictable, inflexible and active after birth.

35

Can a baby detect its mother's mood?

HOW A MOTHER is feeling during pregnancy can also affect her baby's movements. Many readers will be familiar with the opening scene of *The Sound of Music,* in which Julie Andrews skips gaily through alpine meadows belting her heart out. Kazuyuki Shinohara and colleagues at Nagasaki University in Japan showed this uplifting scene to ten women during the last trimester of pregnancy while at the same time using ultrasound to count the number of arm, leg and body movements their babies made. The women also watched a tear-jerking clip from the classic boxing movie *The Champ.* During both movies the women wore earphones, so their babies couldn't hear the soundtrack.

When the women watched the happy film their babies waved their arms, while during the sad movie their movements decreased. Although this study included only a small number of women, this does at least hint that a baby can detect its mother's moods. One way this might occur is through hormones such as adrenaline which are released during times of stress or sadness and can redirect blood flow away from the uterus, which the baby might pick up on.

Janet DiPietro is also studying how foetuses respond to their mother's emotional state. I met her when I was twenty-six weeks pregnant with my first baby, and she allowed me to take part in a study she was running. She has a theory that the bond between mother and baby forms long before birth, and one aspect of this involves the baby tuning in to its mother's emotions. I lay down, and a monitor was wrapped around my belly to detect my baby's heart rate and movements. Electrodes were stuck to my fingers and chest to pick up my heart rate and other physical responses that might be generated, and I was shown a series of videos about giving birth.

First, I watched a happy film in which new mothers talked about their birth stories. I felt warm and excited about my own approaching labour and the baby I would soon meet. Then the film changed, showing graphic scenes of women experiencing difficult births and C-sections which made me want to look away from the screen.

What's interesting is how my unborn baby seemed to react to these scenes. When we looked at the trace of her movements, there was a clear change in her behaviour during the gory video: first she stopped moving as if she was paying attention to what was going on in my body, then she started wriggling all over the place. DiPietro is still analysing the results of this study to see if her theory holds up, but an earlier study did find a connection between a baby's response to its mother's emotions before birth and what the baby was like once it was born. The baby's heart rate generally slowed and its movement was reduced when its mother watched the birth videos, although the heart rates of many babies shot up during the graphic birth scene if the mothers became agitated by it. And those babies that were the most in tune with their mums' emotions also seemed to be more irritable six weeks after birth.

Curiously, the relationship may work both ways, with foetal movements having an unconscious effect on the mother as well – possibly training her to become alert to her child's needs and fall into line with its activity patterns so as to be ready for when it is born. DiPietro has found that every time a foetus moves, its mother's heart beats slightly faster and her 'fight or flight' response is stimulated, even if she isn't conscious of the baby's movements. Although more research is needed to understand the significance of this finding, DiPietro has some ideas. Until recently, the assumption was that the relationship between mum and baby was just one-way, but this suggests that it's a bidirectional relationship, she says. 'The mom's body affects the foetus, but the foetus also affects the mom. I think it's preparing women to pay attention to their infants.'

36

What do babies learn during their time in the womb?

A BABY IS wrenched bleary-eyed and sticky from its mother's body, a blank slate upon which its life and experiences will now become etched. If this is your perception of the moment of birth, think again. A newborn baby has already clocked up a raft of sensations and experiences from its time in the womb.

Two months before a baby is born, it is already behaving pretty much as it will once it enters the wider world. Although it spends around 90 per cent of its time asleep, that sleep is divided up into REM sleep, when its eyes dart back and forth under its eyelids, and non-REM sleep. When its eyes are open it can see (as some light occasionally filters into the womb, bathing it in an orange glow), and for several months it has been able to hear as well. Taste and touch are not neglected, as the unborn baby is treated to a cocktail of flavours from its mum's diet and spends its waking hours exploring its own body and the inner walls of its mother's soft, warm uterus.

Along with these senses comes the ability to remember. Take sounds. From around twenty-two to twenty-four weeks of pregnancy, a baby's ears have developed enough for it to hear noises from the outside world, and the baby may sometimes 'jump' in response to a particularly loud noise. Although many of these sounds will be muffled by the mother's tissue and amniotic fluid, frequencies of around 125–250Hz – the fundamental frequency of the human voice – remain relatively clear, which means that an unborn baby can tune into its mum's voice, as well as to the sound of other noises in her environment. Several studies have inserted microphones into the uteruses of pregnant women, so we know that a baby's world is far from silent. There is the ongoing thump of the mother's heartbeat and the 'swoosh' of blood pumping through her arteries. There are the gurgles of her digestive system and the distorted sounds of the outside world. The mother's own voice is the loudest, as it is conducted through her bones and down into her uterus. Studies have also found that the baby's heart slows when its mother talks, suggesting that it knows her voice and is comforted by it.

This also suggests that unborn babies can remember voices. Several types of memory have been demonstrated in foetuses, from basic habituation (where you learn to filter out irrelevant noises like the ongoing hum of a dishwasher) to classical conditioning (where you learn to associate an object or action with a particular outcome, Pavlov's dogs being the most famous example). Pavlov's dogs were trained to associate a ringing bell with being fed, so, eventually, the mere sound of a bell ringing made them salivate in anticipation of what was to come.

The earliest study to demonstrate that foetuses could learn was undertaken in 1925 by a German paediatrician called Albrecht Peiper. He noticed that a foetus would jump if a motor-car horn was sounded close to its mother's belly during late pregnancy, but after a few beeps it no longer seemed to notice – a form of habituation. More recent studies have shown that a baby can habituate itself to sounds from around twenty-two to twenty-four weeks of pregnancy.

Unborn babies can also learn to associate a piece of music with relaxation, which might prove a practical way of calming your baby after birth. In one study, pregnant women were conditioned to relax when they heard a twelve-second burst of music. After birth, those babies that had listened to the music during pregnancy stopped crying, opened their eyes and made fewer jerky movements when they heard the same piece of music played back to them. Such associations are thought to form after around thirty-two weeks of pregnancy.

In addition, unborn babies seem to remember everyday sounds without being conditioned to respond in a certain way. In one famous experiment, Peter Hepper of Queen's University in Belfast studied a group of women who regularly watched the Australian soap opera *Neighbours* during their pregnancies to see if their babies could learn the theme tune. When they were played the tune several days after birth, they stopped moving, became more alert and their heart rates dropped, suggesting that they remembered it, while babies who had never heard the tune showed no response. None of the babies responded to unfamiliar tunes.

Hepper did this experiment with thirty-week-old foetuses and thirty-seven-week-old foetuses, but only the thirty-seven-week-olds could remember the tune after birth. What's more, if the theme tune was played to these babies twenty-one days after birth, their memory of it had apparently been lost, suggesting that foetal memories may be fairly short-lived.

Researchers at Maastricht University have also been testing the length of time foetuses hold on to memories. They found that if they placed a vibrating wand on a pregnant belly, this generally prompted the baby inside to kick – at least, the first time they felt it. However, babies quickly got used to the vibrations and began to ignore them. The researchers then tested how long it took babies to start responding to the vibrations again, assuming that this would be a measure of their memory of the experience. In thirty-week-old foetuses, it took just ten minutes, but by thirty-four weeks babies seemed to remember the vibrations for at least four weeks.

Although most experts agree that any memories formed in the uterus are likely to be fairly basic, and many animal foetuses have shown similar learning abilities, this doesn't mean that foetal learning isn't important. One theory is that these memories help to ease the transition between the womb and the outside world by teaching babies what sounds are 'safe', so that they aren't alarmed by everyday noises such as the doorbell or the television after birth.

All of this suggests that there may be ways of making your unborn baby's introduction to the wider world smooth. Read it a nursery rhyme, sing it a song, and continue doing the same thing once it is born. If nothing else, it may help bring you closer to the little person you're going to meet in just a matter of weeks.

37

Will playing Mozart to my bump make my baby more intelligent?

FOR SOME AMBITIOUS parents, it's never too early to start trying to boost their baby's academic performance. Several studies have suggested that playing classical music to infants can boost their intelligence; this phenomenon has been dubbed 'the Mozart Effect'. This has led to the development of several 'antenatal universities' – usually programmes of music or natural sounds designed to be played to babies during the final weeks of pregnancy to kick-start their brain development.

This idea started in the 1980s when a researcher called Donald Shetler played classical music to the abdomens of sixteen pregnant women at the Eastman School of Music in Rochester, New York, and noted that this seemed to be associated with the early development of 'highly organized and remarkably articulate speech' once the children were born.

Since then, several studies have hinted that classical music can boost spatial reasoning, or the ability to imagine objects and reason with them (which is important in fields such as architecture and engineering), when played to young children.

In 1998, a study was published showing that rats exposed to different types of music while in the uterus and during the first sixty days of life showed different levels of performance when put into a maze. Those that had listened to Mozart completed the maze more quickly and with fewer errors than rats that had listened to music by modern composer Philip Glass, white noise or silence.

Whether the same applies to humans is less clear. Human

foetuses do show changes in heart rate and movement in response to both music and the human voice. They also seem to remember music or sounds they have been exposed to during their mother's pregnancy (see 36: 'What do babies learn during their time in the womb?'). Evidence that classical music has any lasting effect on human babies remains weak, however. 'There are many claims but as yet no work has been published of peer-reviewed scientific standard to demonstrate they actually have an effect,' says Peter Hepper.

That's not to say that playing music to your unborn baby is harmful. It may relax you, and your baby will probably respond to this. It may also ease your baby's transition into the outside world by providing it with something familiar. And even if Mozart doesn't make your baby brainier, it may have other benefits. In one study, playing Mozart to preterm infants on two consecutive days seemed to boost their weight gain, seemingly by relaxing them and reducing the amount of energy they burned.

38

Will my baby look like Mum or Dad?

MOST PARENTS BELIEVE that their child is the most gorgeous being on the planet, no matter what anyone else thinks. The chances are you'll also rate your partner as pretty attractive. But maybe there are facial features you secretly hope your child isn't landed with – and this probably applies to your own facial features as well. It might be a father's prominent nose or your own thin lips that you'd prefer your baby didn't inherit.

We all inherit some genes from our mother and some from

our father, but several things happen on the way to making a new baby to ensure that he or she is unique. The first happens before the baby is even conceived. Most cells in our bodies contain forty-six chromosomes (twenty-three from our mother and twenty-three from our father), but egg and sperm cells are exceptions, containing just twenty-three chromosomes in total.

When eggs and sperm are made, every chromosome you inherited from your mother matches up with the equivalent chromosome from your father, and they shuffle their genetic information about before being divided between two new cells. This means that some of your mum's genes will end up next to your dad's genes on the same chromosome, and each new egg and sperm cell will have lost around half of your mum's and dad's genes.

However, that's not the only way genetic variation can occur. Once an egg is fertilized, mutations can creep into the embryo's DNA as its cells divide, and although the maternal and paternal copies of most of the 20,000-odd genes the baby inherits will be equally active, in a small number of cases the mother's or father's copy will be switched off. One example of this is the genes that control the formation of the placenta, which is largely guided by the father's genes.

Although most people agree that children tend to take after their parents, pinpointing exactly how physical characteristics such as the size of your nose or the shape of your teeth are inherited isn't easy. Rather than being controlled by a single gene as some diseases are, physical traits are usually the result of lots of different genes working together – often combined with the effects of the physical environment as well.

Even eye colour isn't as simple as it once seemed. People used to believe that variations of a single gene determined your eye colour, with brown eyes being dominant over blue and green. This is still broadly true, but we now know that there are about sixteen genes that can influence eye colour (some more than others). People used to think that it was impossible for two blue-eyed parents to have a child with brown eyes, and if they did, then the mother must have been sleeping with someone else. In fact, because of the complex interplay of all these genes, up to 25 per cent of children with blue-eyed parents may end up with brown eyes, depending on the genes their parents possess.

Hair colour is similarly complex, with one exception: red hair. A gene called MCR1 seems to be the key player in deciding whether or not a child has red hair, and the flame-haired variant is recessive. This means that if a child has one copy of the red-hair variant and one copy of the non-red-hair variant, they will not have red hair (although they will be a carrier, so their children might have red hair – and carriers often have freckles). The only guarantee of a flame-haired child would be if both parents have red hair. But a blond and a black-haired parent could have a 25 per cent chance of a red-haired child if both of them are carriers of a red variant.

The inheritance of facial characteristics such as high cheek-bones or full lips is less well studied, but several research groups have had a go. In one Icelandic study, researchers compared the bone structures of 363 children's faces with those of their parents when the children were aged six and sixteen. The pattern was by no means clear-cut, but boys were generally more likely to resemble their mothers than their fathers, while daughters were influenced by both parents. An earlier Japanese study also noted that sons were less likely to inherit their father's features – particularly their nose shape – while daughters were influenced by both parents.

Although the development of bones and teeth is likely to be a complex interplay between many different genes and environmental factors, these studies hint that several important genes determining bone structure may lie on the X chromosome, which means that sons would only inherit their mother's copies.

39

Do unborn babies dream?

UNBORN BABIES SPEND an enormous amount of their time asleep, but at around thirty weeks this sleep starts to become organized into more adult-like patterns, with four distinct states, from deep sleep to awake. However, there is a key difference: both foetuses and newborns spend far more time in REM sleep than older children and adults, and this is the sleep state associated with dreaming.

Obviously, it's impossible to know what foetuses are dreaming about, and the sort of complex dreams that we experience in adulthood aren't thought to emerge until around five years of age. But unborn babies are immersed in a rich environment of touch, tastes, sounds and movement, so it's quite possible that they are using their REM sleep to rehearse or fine-tune things they've learned during the day. 'Given the foetus's senses are operating and it has the potential to learn, it could be dreaming about the day's experiences,' says Peter Hepper.

REM sleep also seems to be important for the foetus. Certain drugs, including some antidepressants, will suppress REM sleep, and animal studies have suggested that disrupting REM sleep in foetuses results in increased anxiety, disturbed sleep and breathing, and abnormal brain signalling once they become adults. One explanation is that REM sleep may help to perfect the firing of brain cells, ensuring that they are in balance.

Another function of REM sleep might be to train the brain to 'wake up'. The transition between sleep and waking involves the coordination of several brain structures which are also involved in controlling breathing and heart rate, so it is important that these 'learn' to interact properly. REM sleep activates many of these brain structures, so it may be serving as a kind of rehearsal for when the baby has to make these transitions without the life-support system of its mother's blood and oxygen supply.

Weirdly enough, foetal 'yawning' may play a similar role. Babies start to yawn from around eleven weeks and carry on doing so until birth. 'Yawning appears to be not just a matter of opening one's mouth, but a generalized stretching of muscles,' says Olivier Walusinski, a general physician based in Paris who has been investigating this phenomenon. Both foetuses and newborn babies yawn twice as frequently as adults, stretching their mouth muscles once or twice every hour. Yawning also seems to trigger the transition between periods of rest and activity, and, like REM sleep, many of the brain regions involved in waking seem to be stimulated by yawning.

40

Are babies more active at night?

MANY MOTHERS REPORT that their bumps seem particularly prone to kicking when they lie down to sleep at night. Researchers have also noticed this (see 34: 'Does a wriggly bump equal a boister-ous baby?'), although it may be that mums are simply more aware of these movements when they are horizontal because the blows are landing on different internal organs, or because they are not preoccupied with other activities.

As a general rule though, unborn babies seem to synchronize with their mother's circadian rhythm or body clock, so they have some concept of night and day even if no light is reaching their eyes. This may be important in terms of teaching them to sleep and feel hungry at the correct time of day, in preparation for when they are born. Studies in rats have shown that disrupting a mother's body clock during pregnancy results in pups that feed at unusual times of the day. Babies born prematurely seem to have less well-developed circadian rhythms, and there is some evidence that their growth and sleep patterns are disturbed as a result. Several studies have also found that preterm babies whose nurseries are kept dark at night sleep and feed better than those in nurseries where the lights are constantly turned on.

Although such studies are still in their infancy, there may be lessons that all of us can take away from this: if babies are born with a notion of night and day, we should do all we can to reinforce this by giving them plenty of natural light in the daytime and trying to avoid turning the lights on during night feeds. Women should also do their best to sleep at night and get outside during the daytime during late pregnancy (see 100: 'How can I get my baby to sleep through the night?').

41

How do fingerprints develop?

A TEN-WEEK-OLD FOETUS reaches out and brushes against the soft lining of its mother's uterus and as it does so amniotic fluid eddies and swirls around the tiny digits of newly formed fingers and toes, which are already showing the first signs of fingerprints.

For centuries, palm readers have claimed that the lines and creases of the hands and fingers can be used to forecast a person's destiny. They may be closer to the mark than you might imagine, as mounting evidence suggests that fingerprints can provide lasting clues about life in the uterus and may even hint at future risk of disease.

Between the third and fourth month of pregnancy, a baby's delicate skin transforms from a translucent two-layered coating to the more robust and waxy three-layered skin that we have as adults. As this third layer of skin begins to form between the two existing layers it becomes buckled, and ridges begin to form on the skin's surface. These are the beginnings of fingerprints.

Fingerprint patterns are loosely divided into whorls, loops and arches, although there is a lot of variation within these groups. Early in development, raised swellings called volar pads appear on the surface of nascent fingers and toes, which are thought to give rise to blood and connective tissue. Between weeks ten and fifteen these pads stop growing and are absorbed back into the fingertips; precisely when this happens seems to determine the fingerprint pattern you are left with. If the pads are still quite prominent when the third layer of skin begins to form, you get whorls; if they have almost disappeared, you get arches; and if it's somewhere in between, you get loops. These patterns are genetically determined to some extent, but even identical twins show differences in their fingerprints, so some other factor is also clearly at work.

Henry Khan at the US Center for Disease Control and Prevention in Atlanta has been investigating how environmental conditions could influence fingerprint development. Fingerprints were taken from 658 Dutch people born between 1943 and 1947 (some of whom were foetuses at the time of a famine that struck the Netherlands between 1944 and 1945) and the number of ridges on their thumbs and little fingers were counted. Most people have more ridges on their thumbs than on their little fingers, but Khan found a greater difference among babies conceived during summer than those conceived during winter. When they looked at the fingerprints of those who had been foetuses during the famine, this difference couldn't be seen, suggesting that unusually harsh conditions had altered their fingerprint patterns.

In a separate study, Khan looked at whether differences in the number of ridges between thumb and little finger might identify people at higher risk of diabetes. He took fingerprints from 577 people and asked them to take a glucose tolerance test (an early marker of diabetes). As predicted, Khan found that people with normal glucose tolerance had smaller differences in their ridge counts than those with diabetes. Although it's still not clear how a baby's environment could alter its fingerprint pattern, one suggestion is that the availability of nutrients and the presence of stress hormones might interfere in some way.

Birth

Get a Move On

42

Who decides when it's time to come out?

HAVING A BABY is probably the most visceral time of a woman's life. For nine months we are buffeted by hormones that make us sick, tired, emotional and occasionally forgetful. Our bellies swell; our breasts ache; we grow stretch marks and strange new patches of brown skin. Once the baby is born, it is usually placed, slippery and warm, on our chest, and a new tide of hormones overwhelms us, prompting our breasts to leak milk and sending us on another crashing rollercoaster of emotional highs and lows.

The aspect of all this that I found hardest to deal with was the loss of control over my own body – and nowhere was this more acutely felt than in the timing of the actual event itself: birth.

For nine months I diligently counted off the weeks, each one taking me a step closer to the day that had been heavily circled in my diary since my first doctor's appointment: the due date. The two weeks before and all dates after that were a no man's land of uncertainty. Finally, I reached my due date and, hey presto, I was still pregnant. I felt like I was trapped in limbo – impatient and utterly powerless.

The trigger of birth remains one of the biggest mysteries of the human body. For years, people assumed that the baby came out when the mother's uterus could stretch no further and deflated like an overblown balloon, pushing the baby with it. A later view was that labour started when the mother was no longer able to supply enough oxygen and nutrients to sustain the baby's growth and the baby somehow triggered its own delivery. More recently, the focus has shifted, passing the reins of control to neither mother nor baby but to that gelatinous mass that has sustained the baby's growth over nine months: the placenta.

Two hormones control the contractions that thin the cervix and expel the baby from the uterus: oxytocin and prostaglandin. It is these hormones that are given if labour has to be induced artificially or speeded up if it isn't progressing fast enough. But what prompts a woman's body to start producing these hormones in the first place?

This is a complex puzzle which scientists have spent years trying to unpick. Simplified, one of the current leading theories goes like this. Since shortly after conception, a tiny clock has been ticking in the cells that become the placenta, and it grows steadily louder and louder until those ticks can no longer be ignored. The physical manifestation of this clock is a hormone called CRH, which is produced by the placenta throughout pregnancy in larger and larger amounts until it reaches a critical threshold and unleashes a cascade of hormones which ultimately triggers birth.

First, the baby's brain releases two hormones, ACTH and DHEA, which prompt the placenta to start churning out large amounts of oestrogen. This in turn initiates the production of prostaglandin and oxytocin, kick-starting the contractions of labour.

The idea of a placental clock is a nice one, but it doesn't explain everything. For example, what causes the cervix to soften and prompts your waters to break? One idea is that the immune system is involved. As the baby matures, its lungs begin to produce a substance called surfactant which stops their moist linings from sticking together. But surfactant also gets out of the lungs and into the amniotic fluid, where it rallies immune cells to the uterus. These immune cells then release chemicals that soften the cervix and weaken the membranes containing the baby and amniotic fluid, making them more vulnerable to bursting. The fact that premature labour is often associated with infection lends support to the involvement of the immune system, as infection also activates the immune system.

Regardless of how it starts, it's true to say that labour varies massively among women, as does their perception of the pain it causes. Some get stuck in the first stage of labour for days. Others whizz through this and spend most of their labour in the second (pushing) stage. Still others are unable to give birth vaginally and require a C-section. Labour is nothing if not unpredictable. The best advice is to keep an open mind and be prepared for all possible scenarios.

43

Can my state of mind delay labour?

MY FIRST CHILD, Matilda, was due on 31 August – the last day of the UK school calendar – which would have made her the youngest in her year group. Although plenty of geniuses have been born in August, there is some evidence that children who are the oldest in their school year are at an academic advantage, so I couldn't help inwardly crossing my legs and hoping Matilda would hold on for an extra day (she was born a week overdue in the end, which is common for first babies; see 44: 'Is it normal for pregnancies to run past their due date?').

It's widely assumed that women have no control over when they give birth, but a recent study has challenged this. Rebecca Levy and her colleagues at the Yale School of Public Health examined eleven years of US birth records dating from 1996 to 2006, paying particular attention to the two-week periods surrounding Halloween and Valentine's Day. Regardless of the type of delivery, women were 5 per cent more likely to give birth on Valentine's Day and 11.3 per cent less likely to give birth on Halloween than in the two weeks surrounding these days.

As Halloween often carries negative connotations of death and evil, Levy suggests that women may subconsciously want to avoid having a baby on that day, while Valentine's Day, with its associations of cute cherubs and love, may elicit the opposite reaction – although she hasn't yet pinned down a biological mechanism by which positive or negative thoughts could trigger labour.

44

Is it normal for pregnancies to run past their due date?

FOR CENTURIES, the accepted wisdom has been that pregnancy lasts for nine calendar months and that you can calculate your due date by adding seven to the date of your last period then counting back three months. This calculation became known as Naegele's rule, after the nineteenth-century obstetrician Franz Naegele, who first published it in 1812.

It may seem ludicrous that a 200-year-old calculation is still used to predict length of pregnancy today and, indeed, many hospitals now use ultrasound to confirm the expected date of delivery based on a baby's size at around twelve weeks of pregnancy, which is considered more accurate. Studies that have compared ultrasound dating to Naegele's rule have found that, on average, Naegele's date is usually three days too early.

However, this too is an average. First-time mums generally have longer pregnancies than women expecting their second, third or fourth child. One study of Caucasian women found that although 61 per cent of existing mums ran overdue according to Naegele's rule, they were only three days late on average (as ultrasound dating would predict). In contrast, the average first-time mum gave birth eight days later than Naegele would have predicted, and 81 per cent were overdue.

Pregnancy length also varies according to racial background, with more South Asian and black women giving birth before thirty-nine weeks than white women. Their babies also seem to be more developmentally advanced than those of white women born at the same stage.

All this could have medical implications. Most obstetricians believe that remaining pregnant beyond forty-two weeks may put the mother's and baby's health at risk, so they often induce labours that run seven to fourteen days past the expected due date. The main reason they give is the increased risk of stillbirth among babies born after forty-two weeks, which doubles from one per thousand pregnancies at forty-two weeks to two per thousand pregnancies at forty-three weeks. However, if the due dates are out, then women may be being induced unnecessarily, which can carry some risks of its own (see 47: 'Will being induced mean I'm less likely to have a natural birth?'). In industrialized countries, rates of induction range between 10 and 25 per cent of pregnancies, with around 20 per cent of women in England being induced (22 per cent in the US) – although there is an ongoing debate about whether this number is too high.

45

Can curry or anything else help trigger labour?

SEX, CURRY, RASPBERRY-LEAF TEA – there are a multitude of home remedies for kick-starting labour. By the time women reach the fortieth week of pregnancy, they're usually desperate to meet their baby and will try anything to get things moving. A friend of mine resorted to hours of frantic nipple-tweaking, which left her with very sore breasts, but she went into labour that night. Heavily pregnant myself, I tried imitating my friend, but got nothing but a minor twinge to show for it. So, pushing my scepticism aside, I paid a Chinese therapist £30 for a massage and, eight hours later, my contractions started. An effective therapy, or was it simply that I was already seven days overdue and labour was destined to start anyway?

Fortunately, we don't have to rely on anecdote to find out whether such home remedies really work, as a number of scientific studies have been done to evaluate them.

Raspberry-leaf tea

For centuries, herbalists have been using the leaves of the common red raspberry (Rubus idaeus) to prepare the uterus, and a recent survey of US-based midwives found that 63 per cent of them recommend it for inducing labour. However, in studies where raspberry-leaf extract has been given to animals or added to strips of uterine tissue, the results have been mixed: some showed that it relaxed the muscle, others that it caused more coordinated contractions, while still others suggested that it increased the length of pregnancy.

Even if raspberry-leaf tea does affect the uterus, it's not clear how much of the stuff you'd have to drink for the active ingredients to work their magic. One small study, which compared women who took raspberry-leaf tablets twice a day from thirty-two weeks with women who didn't, found a very small reduction in the length of the second (pushing) stage of labour in the women who took the tablets, but the effect didn't reach statistical significance, meaning that it could have been down to chance.

Curry or castor oil

Disappointingly, there have been no studies to assess the value of curry in initiating labour (and I would happily volunteer myself as a guinea pig if there were). The theory is that the spice in curry irritates the bowels and triggers muscular contractions that send you rushing for the bathroom, but may also spread to the womb and initiate labour.

Better studied is castor oil, which is thought to have a similar effect. Castor oil has been used to induce labour since Ancient Egyptian times and was still being recommended by obstetricians in Western countries until the 1950s, when newer drugs such as syntocinon (a synthetic form of oxytocin) became available. It is still used in many developing countries today. Apparently, it's pretty unpleasant to drink, and may make you sick, as it has the taste and texture of very thick cooking oil. In 2012, an Israeli group announced the results of a prospective, randomized, double-blind, placebo-controlled study (which is about as good as it gets) on whether drinking castor oil induced labour: it found that the odds of entering labour within twelve hours of drinking castor oil were three times higher than if you drank sunflower oil instead. Women also seemed to progress through the first stage of labour more quickly. Unfortunately, there were just eighty women in the study, so more research is needed. Whether curry is an effective substitute, we don't know, but it would probably have to be spicy enough to give you diarrhoea – in which case you might prefer to try castor oil. The usual dose is 114ml (4fl oz), which can be chased down with orange juice, although it may be wise to speak to your doctor or midwife before trying it.

Pineapple

Pineapple contains an enzyme called bromelain, which could theoretically help to soften the cervix – although whether it would survive the harsh conditions of the digestive system and get into the blood in high enough amounts to have any effect seems unlikely. Tasty as it is, there is no scientific evidence to support the use of pineapple in inducing labour.

Nipple stimulation

Tweaking, massaging or pumping the breasts all spark the release of oxytocin, which is one of the hormones that causes labour contractions. A review of six trials involving a total of 719 heavily pregnant women found that more than a third of women who stimulated their breasts were in labour within seventy-two hours, compared to just 6 per cent of those who didn't. The amount of time the women spent tweaking their nipples or pumping their breasts varied from a total of one to three hours a day (with rests in between), and the women generally switched breasts every ten to fifteen minutes.

Although some doctors fear that nipple stimulation might overstimulate the uterus and potentially upset the baby, the review found no evidence of this (although its authors suggest avoiding nipple stimulation in high-risk pregnancies, just to be on the safe side). Nipple stimulation also seemed to reduce the risk of excessive bleeding after birth, presumably because the extra oxytocin caused the woman's uterus to contract more strongly and close off the blood vessels feeding the placenta after birth. A shot of oxytocin is often offered to birthing women in order to achieve the same effect.

Sex

Although sex may be the last thing on your mind when you're about to give birth, many people will tell you that semen contains prostaglandins, which help to ripen the cervix. Possibly this rumour was started by men in a last-ditch effort to get some action before the baby arrives, but it seemed logical enough to warrant further investigation.

Swedish scientists recently asked twenty-eight heavily pregnant women either to have unprotected sex 'with vaginal semen deposit' three nights in a row or to avoid sex completely. Because breast stimulation might have an independent effect, both groups were told that the nipples were absolute no-go areas. In this admittedly small study, sex didn't seem to do anything. There was no difference between the women in terms of the baby's position, the softness of the cervix or the number of women who gave birth within the next three days. Having sex is unlikely to harm the baby, but it's also unlikely to hasten its birth.

Acupuncture

A handful of small trials have suggested that women receiving acupuncture are less likely to be medically induced, but because these women knew whether or not they were receiving genuine acupuncture, there may have been a placebo effect at work. Many other studies have found no effect.

Homeopathy

The Cochrane Collaboration recently reviewed two trials of homeopathy in inducing labour. One found that women taking a mixture of black cohosh (Actea racemosa), arnica, caulophyllum, pulsatilla and gelsemium twice a day from thirty-six weeks of pregnancy spent an average 5.1 hours in labour, compared to 8.5 hours for those who took a placebo pill, and suffered fewer complications – although the authors of the review said the study was too small to draw any firm conclusions. The second study looked at women whose waters had broken prematurely and found that those taking an hourly dose of caulophyllum took an average thirteen hours to begin regular contractions, compared to 13.4 hours in women taking no supplements – a small enough difference for it to be down to chance.

46

Does having a membrane sweep work?

IF YOU SAIL past your due date, you might be offered something called a membrane sweep. It's certainly not the most dignified or comfortable procedure, involving a doctor or midwife sticking her fingers into your cervix and trying to stretch it and (if she can reach that far) separate the membranes containing the baby from the top of the cervix. This supposedly releases prostaglandins that further soften and thin the cervix, hastening the onset of labour.

Just how successful your midwife will be depends on the state of your cervix when she carries out the sweep. At forty to forty-one weeks, some women will already have started to dilate, while others may be experiencing contractions even though their cervix shows little sign of softening, and the midwife may struggle even to reach it.

Plenty of studies have investigated whether membrane sweeping works, with mixed results. For this reason, the Cochrane Collaboration tried to review as many as it could in an attempt to draw some firm conclusions about the procedure. Generally speaking, membrane sweeping does increase your chances of going into labour within the next forty-eight hours or giving birth within the next week if you are already overdue – although it's no guarantee. Eight women would need to have a membrane sweep in order for just one of them to avoid medical induction of labour, and it's hard to know if you'd be in the lucky minority. Having a membrane sweep at thirty-eight to forty weeks of pregnancy also slightly reduces the chances that you'll still be pregnant at forty weeks.

It's a fairly unpleasant procedure and may trigger bleeding

and painful contractions which don't result in labour. In one of the studies reviewed, 70 per cent of women reported significant discomfort, while a third reported significant pain. For this reason, the review's authors say the rationale for membrane-sweeping healthy women who aren't overdue is questionable.

47

Will being induced mean I'm less likely to have a natural birth?

A WOMAN SAILS through pregnancy healthy, happy and excited about meeting her baby for the first time. As the due date passes, frustration and impatience often set in, but for women who stubbornly remain pregnant once they pass the forty-one-week barrier, a new terror awaits: the threat of induction.

Supporters of natural childbirth sometimes warn that being induced will make contractions more painful and difficult to cope with, which means that you're more likely to need an epidural, which in turn can make a C-section or instrumental delivery (involving forceps or a ventouse) more likely. What's more, because induction is a medical procedure, your baby's heart rate will need to be continuously monitored, which makes it harder to move around during labour or use a birthing pool.

The evidence suggests that many of these fears are overblown (see also 66: 'Does an epidural make a C-section more likely?'). Medical induction is usually recommended once women reach forty-one or forty-two weeks, and several drugs can be used. The first strategy is usually to give prostaglandins to help soften the cervix and make it easier to open. If this doesn't help, a plastic tube is usually inserted into the back of the hand to provide a drip containing syntocinon – an artificial version of the hormone oxytocin, which regulates contractions.

Some studies have suggested that women who are induced tend to request an epidural sooner than those entering labour naturally, which might suggest that the contractions are more difficult to cope with. However, it's also possible that because induction is generally carried out in a hospital ward under a doctor's supervision, women are simply more likely to be offered an epidural rather than being encouraged to use other methods of pain relief.

So what about the claim that being induced increases your risk of needing an emergency C-section? So widespread is this dogma that in 2009 the US Agency for Healthcare Research and Quality (AHRQ), which produces guidance for doctors, decided to take a closer look. Although some studies have suggested that induction boosts the risk of having a C-section, they have tended to compare women who are induced to women who enter labour spontaneously. This might seem reasonable, except that most women who are induced are overdue, so it would be better to compare them with other overdue women who are simply wait-ing and monitoring the situation. These women may eventually enter labour spontaneously, or they may have to be induced – but watchful waiting can bring problems of its own, such as ending up with an excessively large baby that needs to be delivered by C-section anyway.

After reviewing seventy-six studies, the AHRQ found that women who were induced were actually 20 per cent less likely to need a C-section than women who simply watched and waited. Their babies were also around 50 per cent less likely to have meconium (poo) in their amniotic fluid, which is a sign of maturity and is more common in post-date babies, but can increase the risk of breathing problems. Most of the studies analysed looked at women at or beyond forty-one weeks of pregnancy. Before this, the risk of C-section appeared to be more or less equal in both groups of women. Two other large reviews published within the last ten years reached similar conclusions.

Women may also fear that if being induced makes an epidural more likely, then this may boost the risk of an instrumental delivery (one involving forceps or a ventouse). A separate review of nineteen trials found a slightly increased risk of instrumental delivery among women who were thirty-seven to forty weeks pregnant when they were induced, but this disappeared when one looked at women who were forty-one to forty-two weeks pregnant. This is important because most women who worry about being induced will fall into the second category. Bear in mind that if your pregnancy is overdue and you aren't induced then your baby will keep on growing and may therefore be harder to push out when the time finally does arrive. A similar picture emerges if one looks at women who are offered a syntocinon drip because their labour slows or stalls. Overall, this seems to shorten labour and ultimately make women less likely to need a C-section or instrumental delivery.

The message I take from all of this is that, although being induced may quash your dreams of a home or midwife-led birth, if you're still pregnant two weeks past your due date then it may be worth listening to the advice of your doctors. You can still hope for an otherwise uncomplicated natural vaginal delivery, and it may ultimately be safer for both you and the baby.

48

Will rocking on my hands and knees cause my baby to turn over?

BY THE END of pregnancy, 90 per cent of babies are head down and two-thirds of them are facing towards their mother's spine, which is the optimum position for fitting through her pelvis. However, a stubborn few seem determined to face the other way around, which makes birth more painful, and a breech baby who remains head up will often need to be delivered by C-section.

Because the head is the heaviest part of the body, people may tell you to get down on your hands and knees and use gravity to turn a difficult baby. Unfortunately, there is little published evidence to suggest that this works. One recent review of three studies by the Cochrane Collaboration found that although getting on to your hands and knees for ten minutes a day caused the baby to rotate in the short term, babies didn't stay that way for long. Several other studies suggest that, although babies may shift their position according to what the mother is doing at the time – aligning their bodies with the mother's when she rests, for instance – they often revert to their original position or a different position afterwards.

Women shouldn't give up on the hands-and-knees position entirely, though: the Cochrane review found that it was effective at relieving back pain during labour.

Midwives may also try using something called external cephalic version (ECV) to turn a breech baby, which involves prodding and pushing at your bump in order to encourage the baby to flip a somersault. Success rates for this technique range from 35 to 57 per cent for women in their first pregnancy, and from 52 to 84 per cent for existing mothers, so it seems worth a try.

49

Can I take a bath once my waters have broken?

RELAXING IN WARM water is a great way of staving off early labour pains, but women are sometimes told that this is dangerous once your waters have broken because bacteria in the bathwater could be flushed into the uterus and infect the baby. This rumour seems related to a common distrust of water births, which were also once considered dangerous because of fears of infection.

However, when Swiss researchers compared the risk of infection in mothers whose waters had ruptured, they found no difference between women who took a warm bath and those who didn't.

Several studies have also looked at infection rates in women who choose to have water births and found no increased risk of infection in either mother or baby. Indeed, despite water often becoming contaminated with faeces during a water birth, some obstetricians believe this may even help to protect newborns against subsequent diseases and allergies (see 54: 'Are vaginal births really better than C-sections?').

The Big Push

50

Why do humans find it so difficult to give birth?

WATCH A DOG or a horse giving birth and their babies just seem to slip out, with none of the pain and screaming associated with human labour. During my first pregnancy, I fell for a book which told me that, because animals have relatively pain-free births, women could have a similar experience if only they banished fear. Having gone through labour, I now think this theory does women a great injustice, because it implies that requesting pain relief means you have somehow failed. Research highlights several reasons why humans don't give birth like animals – and none of them have anything to do with eliminating fear.

In the grand scheme of things, we humans have existed only for a wisp of time. Although the first mammals appeared 100 million years ago when dinosaurs still roamed the Earth, our early ancestors began walking on two legs only around 3 million years ago. This transition from four legs to two required that the human pelvis became narrower, but since then its size has remained pretty much unchanged, while the human brain has almost quadrupled in size. 'Human labour is an uneasy compromise between our need to run and our need to think,' says Philip Steer, professor of obstetrics at Chelsea and Westminster Hospital in London. 'We're erect, bipedal animals, but we have very big brains.'

A sheep or rat's uterus is paper-thin compared to that of a human, because a lamb or rat pup meets little resistance on its journey into the wider world. In contrast, the human uterus is a powerful muscle that can still take hours to push the baby out.

Even some of our closest living relatives have an easier time of it than we do. Imagine a hard-boiled egg cut in half, with the yellow yolk surrounded by a thick rim of egg white. If a baby gorilla's head is the egg yolk, then the egg white is similar in proportion to the amount of space a baby gorilla's head has when it passes through its mother's pelvis. Although it has to push against muscle and soft tissue, there is no bony pelvis standing in its path.

Now imagine trying to squeeze a golf ball through an egg-shaped hole: it barely fits. That's how a human head looks in proportion to a human pelvis. All those brain areas we've developed that give us personality, and the ability to forward-plan and socialize, come at the expense of making it harder to squeeze the head out at birth. We've adapted to some degree, as the bones of the human head will overlap when squeezed (which is why many babies born vaginally have 'cone heads' to start with), but it's far from easy.

51

Why are human babies born so helpless?

THAT SAME CHALLENGE of, effectively, squeezing a pinhead through the eye of a needle also helps to explain why human babies are born so helpless. Human pregnancies last an average of forty weeks, which is similar to those of chimpanzees or gorillas, but human babies are born premature by comparison. Besides not being able to walk until they're around a year old, a newborn human has a brain that is just a quarter of the adult size, compared to half the adult size for baby chimps and gorillas. This is probably so that the head remains small enough to fit through the birth canal. 'The joke is that if humans were born as mature as chimpanzees, human pregnancy would have to last seventeen months,' says Steer. 'Think of the complaints we would get in the antenatal clinic then!'

In addition, human babies have to twist their way through the birth canal in order to get out, which means that they usually come out with their head facing towards their mother's bottom – unlike other apes, whose babies are born face up. This may be why midwives exist in virtually every human society that has been studied. Although other apes see their baby's face emerge, can clear any mucus from its mouth and nostrils and remove the umbilical cord from around its neck, it's simply not practical for humans to give birth on their own.

52

What's the biggest baby that's ever been born?

ACCORDING TO *The Guinness Book of World Records*, the largest baby ever born was a boy weighing 10.8kg (23lb 12oz) and measuring 76cm (30in) in length. That's around three times the average weight of 3.4kg (7lb 8oz), and 25cm (10in) longer than average. The boy was born to a giantess called Anna Bates, who herself stood 2.27m (7ft 5.5in) tall, in Seville, Ohio, on 19 January 1879. Sadly, the baby died eleven hours later.

The heaviest baby born to a normal-sized mother and father was a boy weighing 10.2kg (22lb 8oz), who was born to Carmelina Fedele in Aversa, Italy, in September 1955. More recently, an Indonesian woman gave birth by C-section to a boy weighing 8.7kg (19lb 3oz) in 2009. The mother is diabetic, so high sugar levels in her blood may have fuelled the baby's growth.

The title for smallest baby goes to Amillia Sonja Taylor of Florida, who was born weighing just 283g (10oz), and survived, despite being born extremely premature – at twenty-one weeks and six days of gestation.

As for the largest number of children born to one woman, Fyodor Vassilet gave birth to sixty-nine children during her lifetime. Born in Russia in 1707, she experienced twenty-seven pregnancies, giving birth to sixteen sets of twins, seven sets of triplets and four sets of quadruplets, before dying at the age of forty.

53

Is there any way to predict how long labour will last?

WHEN I WAS expecting my first baby, I was told that I should look to my mother's experience of labour for an idea of how my own would progress. She said it took around nine hours from her first contraction to my slithering into the world. So when my own contractions started at 1 a.m. on a Monday morning, I diligently switched on my TENS machine and fully anticipated that I would have my baby in my arms by lunchtime.

Twenty-four hours later, I was in hospital being told that, despite my contractions having been coming every three minutes for the past six hours, my cervix had dilated only three centimetres and I was barely in 'active labour'. It wasn't until 7.30 p.m. on Tuesday that my daughter finally burst into the world – some thirty-two-and-a-half hours late by my original calculations.

The average length of labour for a first-time mum is somewhere between thirteen and seventeen hours and can be divided into three stages. The first stage is when the cervix is thinning and dilating to the point where it is open enough to let the baby's head through, which typically takes twelve to fourteen hours in first-time mums (although this first stage can be further subdivided into a latent phase, when contractions are occurring but the cervix isn't dilating, and an active phase, when it is steadily dilating). Next is the second or 'pushing' stage, which typically takes one to two hours, followed by the third stage, which involves expelling the placenta and takes up to an hour. It is true that subsequent labours tend to be faster than the first – around six to eight hours for the first stage of labour and five to sixty minutes for the second stage.

Several things are known to influence how long or difficult a woman's labour will be. In terms of the first stage, women whose baby's head has not yet started pressing on the cervix (something called engagement) tend to have longer latent phases. Weirdly, this is also true of women who have had more than five babies, presumably because their uteruses are weaker and less able to push the baby against the cervix (although the second stage of labour tends to be much faster in these women). Obese mums also tend to have longer first stages of labour, although the reason for this is unclear.

More is known about the factors that can slow down the second (or pushing) stage of labour. An obvious one is the size of the baby, as bigger babies tend to take longer to push out and are more likely to get stuck. The size of your pelvis is also important, but just because you have big hips doesn't necessarily mean you'll have an easier labour, because the shape matters too. As a general rule, Afro-Caribbean women tend to have narrower birth canals and are more likely to have trouble delivering if their babies are overdue. This is offset by the fact that they also tend to have shorter pregnancies.

One common myth is that the size of your feet can be used to predict how easy your labour will be. Back in 1988, Scottish researchers examined the association between height, shoe size and the outcome of labour in 563 first-time mums. They found no significant difference in women's shoe size and the rate of C-section, whether they had to be induced, or the length of labour – although the babies of women with larger feet tended to be slightly heavier. However, when it came to height, they found that women who were shorter than 1.6m were more likely to need a C-section because of difficulties with the baby's head fitting through the pelvis. Even so, 80 per cent of women shorter than this still managed to have a vaginal birth.

Another important factor is the position of the baby as it enters the birth canal, which can affect the chances of it getting stuck. Babies whose heads are facing towards their mother's bellies as they enter the birth canal are harder to squeeze out, as they will need to make more twists and turns during their journey.

Sadly, none of these factors are easy to control, or even possible to predict on an individual basis. But there are some steps you can take to increase the chances of your own labour progressing smoothly. Several studies have suggested that women who eat normally in the twenty-four hours before labour and get a decent amount of rest and sleep are more likely to have short labours than those who are tired and hungry. So be sure to keep your energy levels up, drink plenty of water and (if possible) sleep during the early stages of labour, as this extra energy should help when things get tougher.

54

Are vaginal births really better than C-sections?

THE MEDIA HAS long lambasted women who opt to have a C-section rather than pushing their baby out the natural way. In the UK, they're labelled 'too posh to push', the assumption being that women choose to have a C-section either for reasons of convenience or because they are too afraid to go through labour.

Not only are these assumptions insulting to the majority of

women who have a C-section for perfectly valid reasons, they are also largely incorrect. Caesareans are on the rise in many countries, and currently around a quarter of British women give birth by C-section (31.8 per cent in the US), yet there's little evidence that more than a handful are doing it for non-medical reasons. An audit of 32,082 C-sections performed in the UK found that just 7 per cent were performed at the mother's request, and, of these, it's not clear how many involved women with underlying medical problems or those who had previously given birth by C-section. When women expecting their first baby were asked, just 3.3 per cent expressed a preference for a C-section.

Even if large numbers of women aren't pushing to have a C-section, a growing number of doctors believe they should be allowed to choose one if they so desire. In 2011, the UK National Institute for Health and Clinical Excellence (NICE) published guidance stating that all women should have the right to a C-section, even if there is no medical need. The main justification is that, as technologies have improved, C-sections have become far safer – particularly if they are planned, rather than performed as an emergency procedure.

There are many myths and misconceptions about the relative benefits of natural birth versus a C-section, and having experienced both types of birth, I can assure you that there are up-sides and down-sides to both, but let's take a look at the evidence:

A vaginal birth is harder on your sex life

Many pregnant women worry whether their sex life will return to normal after giving birth, and it's no reassurance that up to 24 per cent of obstetricians and 45.5 per cent of urogynaecologists say that they would choose a C-section for themselves or their partner.

The main reason is the higher risk of incontinence associated with vaginal delivery, although nearly 60 per cent add that sexual dysfunction is another reason. But remember: obstetricians and urogynaecologists only tend to see the complicated births and worst cases of birth trauma in their clinics. Just 4.4 per cent of midwives say that they would opt for a C-section – testimony that the majority of women deliver their babies without complications and recover well afterwards.

In the short term, there's no denying that having a baby will probably put a dent in your sex life – regardless of how you give birth (see 72: 'Is it normal to go off sex after birth?'). Up to 86 per cent of women report sexual problems in the first few months after giving birth, and this is more likely among women who need an instrumental delivery (involving forceps or a ventouse) or who tear during birth. Such women often report pain in the area between the vagina and back passage in the eight weeks after birth, and sex may be uncomfortable for up to a year. Even a relatively minor tear or an episiotomy (a deliberate cut to help get the baby out), can be expected to lower your libido and make sex a little uncomfortable for the first three months.

You might assume that a C-section would avoid any such sexual problems, but when it comes to comparing straightforward vaginal births with C-sections, the picture is less clear. Although vaginal births carry a slightly increased risk of damage to the pelvic-floor muscles and the nerves connecting the vaginal area, one large study that compared planned C-sections with un-complicated vaginal deliveries found that 18 per cent of women who had a C-section reported pain during sex in the first three months, compared with 10 per cent of women who gave birth naturally. Having a C-section involves cutting through the walls of the abdomen and uterus, so it's fairly normal to experience some generalized pain in that area. If you have a C-section, you are, however, more likely to have resumed sexual activity a month after birth.

The good news is that, a year on, most women have recovered their sex lives – regardless of how they delivered their baby. When 524 Dutch women were interviewed a year after giving birth, the majority said that they were sexually satisfied, and this was unrelated to the method by which they gave birth.

Vaginal births increase the risk of incontinence

One of the biggest fears surrounding vaginal birth is that it will increase your risk of incontinence. It is true that rates of urinary incontinence are lower in women who have had a C-section, compared to a vaginal birth, although it's no guarantee. This is because much of the weakening of the pelvic-floor muscles associated with motherhood comes as a result of hormonal changes during the early months of pregnancy. Scarring from a C-section can also put pressure on the bladder, making stress incontinence (where you may leak some urine when you cough, sneeze or exercise) more common.

Estimates of the risk vary according to when and how you ask the question, but when a Canadian group recently combined the results of eighteen studies, they found that in the three months to two years after birth, the risk of stress incontinence was 16 to 22 per cent among women who had vaginal births, compared to 10 per cent for women who delivered by C-section. This would mean that ten to fifteen C-sections would have to be performed in order to prevent one case of stress incontinence. When the most serious cases of incontinence were looked at, there was very little difference between the two groups.

It takes longer to recover after a C-section

Women are often told that having a C-section puts them at increased risk of infection, postnatal depression, future infertility and a longer hospital stay. However, the UK's National Institute for Health and Clinical Excellence (NICE) recently reviewed the literature and concluded that for many of these claims the evidence was of very low quality. These risks also have to be weighed against the fact that women who have had a C-section generally describe a better birthing experience both in the immediate aftermath of the birth and three months later.

C-sections are associated with a longer hospital stay, but it's not much longer: an average three to four days, compared with one to two days for a vaginal birth. And if a woman who has had a C-section hasn't suffered any complications and seems fit and healthy, NICE recommends sending her home after twenty-four hours – albeit with painkillers.

Although it may take slightly longer to get back to your pre-pregnancy activity levels after a C-section, some women who have natural deliveries also take time to recover. When nearly a thousand American women were interviewed seven weeks after birth, 65 per cent of those who had a vaginal birth reported no limitations in terms of running, lifting heavy objects and doing strenuous sports, compared to 45 per cent of women who had a C-section, while 90 per cent of those who had a vaginal birth reported no limitations when lifting or carrying bags of groceries, compared to 80 per cent of those who had a C-section.

It's true that having a C-section constitutes major abdominal surgery and therefore increases the risk of complications such as infection, blood clots or blood loss, but there are steps that can be taken to reduce these risks. There is now strong evidence that women should be offered antibiotics before surgery in order to reduce the risk of infection, and they should be monitored for signs of blood clots. Women should take good care of their wound, showering or bathing daily, gently drying the wound afterwards, and being vigilant for any signs of infection such as increased pain, and any discharge, redness or signs of the wound pulling apart. The same applies to women who tear during a vaginal birth and are also at risk of infection.

When Italian researchers compared rates of complications in women undergoing planned and emergency C-sections and vaginal births they found little difference in terms of the risk of infection or serious blood loss between the groups – apart from women who needed an instrumental vaginal delivery and were at greater risk of a serious tear.

C-sections make it harder to bond with your baby

The seeds of this idea can be traced to a speech given by the French obstetrician Michel Odent in 2006, who warned that both C-sections and being induced interfere with the natural production of oxytocin, which helps create a loving bond between mother and baby. Putting aside the fact that one of the drugs used to induce labour is a synthetic version of oxytocin, and that women undergoing emergency C-sections have usually been exposed to plenty of oxytocin (which triggers contractions) before being rushed in for surgery, there is scant evidence to suggest that a lack of oxytocin during birth has any long-term effect on maternal bonding.

One study from 1986 found that women who delivered by C-section handled their babies less frequently a week after birth than mothers who delivered vaginally, but the authors admit that this may be because of tiredness and discomfort, and they didn't check to see how mum and baby interacted several weeks or months down the line.

Similarly, a study published in 2008 which suggested that C-section mums were less responsive to their baby's cries than mums who delivered vaginally drew its results from just twelve women, and used brain imaging to look at how areas related to empathy and reward were activated, rather than measuring hormone levels or seeing how the mothers reacted to their babies in everyday life.

One thing we know about oxytocin is that it is produced throughout the latter stages of pregnancy and is released in response to breastfeeding and simply cuddling and handling your infant, so you should get plenty of it in the aftermath of birth. 'Especially if the baby is breastfeeding, the mum's brain will produce oxytocin and promote bonding irrespective of whether she got synthetic oxytocin or had a C-section,' says Paul Zak, an oxytocin expert at Claremont Graduate University in California and author of *The Moral Molecule*.

Although it is difficult to measure scientifically, some studies have at least hinted that early skin-to-skin contact can help the mother bond with her baby. For example, when the mothers of thirty-two babies who touched or licked the nipple within thirty minutes of birth were compared with those of twenty-five babies whose first contact with the breast was around eight hours after birth, the researchers noticed that women who had experienced early skin-to-skin contact spent more time with their babies, talked to them more during breastfeeding and had lower levels of a stress hormone during breastfeeding over the next four days than mothers who initiated breastfeeding slightly later.

And there are steps you can take to try to make a C-section

feel less clinical and more like a 'natural' birth, which may in turn help to promote bonding. In its latest guidelines, the UK's National Institute for Health and Clinical Excellence recommends that women undergoing a C-section should be allowed to listen to music in the operating theatre, dim the lights, request silence so their voice is the first the child hears – even have the screen lowered at the moment of birth so the mother can watch her baby being born. Assuming there are no complications, they should also be given early skin-to-skin contact if they want it, as this can boost the chances of successful breastfeeding.

C-sections make breastfeeding more difficult

A number of studies have found that women who deliver by C-section are less likely to breastfeed than those who have a natural birth. Why should this be? One clue lies in the fact that women who have had a C-section often take longer to put their baby to the breast than women who deliver vaginally. Mother and baby are usually monitored for several hours following a caesarean birth, and in many hospitals this takes place in separate rooms, which means that women don't get the chance to breastfeed and bond with their baby straight away.

Italian researchers recently compared breastfeeding patterns between women who delivered vaginally, by planned C-section, and by emergency C-section. Upon discharge from hospital, 88 per cent of women who had a vaginal birth were exclusively breastfeeding, compared to 73 to 74 per cent of women who had a C-section. When the researchers examined how long it took for the baby to be offered its first breastfeed, they found it took ten to thirteen hours for those babies delivered by C-section, compared to three hours for babies delivered vaginally. Many researchers believe that such delays interfere with a baby's natural impulse to seek out the breast and suckle in the immediate aftermath of birth, before they get too sleepy. It could also interfere with the onset of milk production in the mother's body, making it harder for breastfeeding to get started.

However, hospitals are increasingly waking up to the idea that women who have had a C-section might want to enjoy immediate skin-to-skin contact with their baby (see 83: 'Does skin-to-skin contact really soothe my baby?'). This is particularly true of planned C-sections and when women remain awake during the operation. In a recent pilot study at San Francisco General Hospital, the proportion of women getting skin-to-skin contact with their babies within ninety minutes of a C-section was increased from 20 per cent to 68 per cent thanks to an awareness drive among medical teams, which in turn boosted the rate of breastfeeding. Among babies who didn't experience skin-to-skin contact within the first ninety minutes of life, 74 per cent needed additional formula milk in the next few days, while only 30 per cent of those that got early skin-to-skin contact needed extra milk.

'Skin-to-skin contact has historically been difficult to implement after caesarean birth due to common hospital procedures,' says Kristina Hung, who led the study. 'However, it is definitely something a woman can request on a birth plan, and it's something I would personally insist upon.'

Women who have had a C-section may need extra help lifting and positioning their baby during the early days, and many breastfeeding experts recommend either lying on your side to breastfeed, or using a football hold, where the baby is tucked under your arm with its legs dangling behind you in order to protect the scar. Women can also breastfeed in a reclined position, with their babies lying on top of them. Babies born by C-section are sometimes sleepier, or less physically developed (if born prematurely), but with the right support there's no reason why they can't be breastfed.

Babies born by C-section are less healthy

There are situations where babies born by C-section are almost certainly healthier. For example, if a baby is breech, its chances of dying or suffering serious injury during birth are around 5 per cent if delivered vaginally, compared to 1.6 per cent if it is delivered by C-section. However, babies born by C-section are at a slightly increased risk of breathing problems in the immediate aftermath of birth – although this risk drops sharply if the C-section is performed after thirty-nine weeks of pregnancy, when the lungs are more mature. Babies born by C-section are also marginally more likely to need resuscitating after birth, while babies born vaginally have a slightly increased risk of rare brain bleeds. Overall, there is no difference in short-term survival rates between babies born by C-section and those delivered naturally.

What about longer term health? A number of studies have suggested that babies born by C-section are at greater risk of asthma and allergy when they get older. When a British group recently combined the results of twenty-three such studies, they concluded that C-section increased a child's risk of asthma by around 20 per cent. This means that if nine out of a hundred children in the general population develop asthma, then approximately eleven in a hundred children born by C-section will become asthmatic. A separate review of twenty-six studies reached a similar conclusion about the risk of hay fever in babies born by C-section and found a 30 per cent increase in the risk of food allergy.

However, this doesn't necessarily mean that C-sections cause conditions such as asthma. For example, premature babies and those with very low birth weights are also thought to be at increased risk of asthma, and these are often delivered by C-section. Breastfeeding is also thought to afford some protection against asthma and allergy, and fewer mums who deliver by C-section tend to breastfeed.

There is, however, one reason to think that babies delivered vaginally might be at a slight advantage. Natural birth is a messy business, and a newborn usually swallows a hefty dose of the bacteria that become smeared across its face, but this rarely causes illness, because it has already received antibodies against many of the thousands of bugs dwelling inside the mother's body. These bacteria then go on to colonize the baby's own gut and help protect against infection, and possibly allergic disease as well.

By contrast, a baby born by C-section never receives this initial inoculation by its mum's bacteria. Instead, the first bugs it encounters may come from the hands of a midwife or the towel in which it is swaddled. Such babies have been shown to have different gut flora to babies born vaginally, and there are some hints that this may leave them more vulnerable to asthma or allergies.

There may yet be a solution to this problem, but it's not for the squeamish: some obstetricians have suggested taking a swab from the mother's vagina immediately after birth and rubbing it over the baby's mouth. This is probably best done only after consultation with your doctor!

55

Are home births more risky than hospital births?

IN 2010, A STUDY by Joseph Wax and colleagues at the Maine Medical Center in the US generated headlines on both sides of the Atlantic with its finding that, although women who give birth at home suffer fewer complications, the risk of their babies dying is trebled.

Even within the scientific journal where it was published, the study led to a flurry of angry letters from other researchers who disputed the conclusions. Wax responded by pointing out that in low-risk women who are cared for by highly trained midwives, planned home births actually reduce infant deaths. In other words, home births are safe, so long as the right women are having them and they are cared for by properly trained individuals. Unfortunately, the reputation of women who choose to have home births as being irresponsible mothers has largely stuck.

Currently, around one in fifty women has a home birth in the UK, compared with around one in two hundred in the US. One reason for this disparity is that the American College of Obstetricians and Gynecologists remains concerned about home births, citing a lack of good-quality evidence to prove their safety, whereas both the UK's Royal College of Midwives (RCM) and the Royal College of Obstetricians and Gynaecologists (RCOG) support home birth for women with uncomplicated pregnancies.

Wax's 2010 study is one of the largest to tackle the issue of home birth, so it's worth taking a closer look at its conclusions. The study was actually a review, pulling together the results of twelve separate studies and analysing a total of 342,056 planned home births and 207,551 planned hospital births within the UK, US, Canada, Australia and several European countries.

As you might expect, home births were associated with fewer medical interventions, including epidurals, electronic heart-rate monitoring of the baby, C-sections, instrumental deliveries and episiotomies, as many of these things are difficult to do at home. Even so, mothers who had home births experienced fewer infections, tears and serious bleeds, compared to mothers who delivered in hospital. This doesn't necessarily mean home births are better for women: the study included existing mothers, and if they had experienced complications during their first birth, they may have chosen a hospital birth this time around, while women expecting large babies (who are more likely to tear) would also probably have opted for a hospital delivery. Similarly, when it came to the babies born at home, these were less likely to be premature, of low birth weight or to need additional help breathing than those born in hospital – many such babies would be considered high risk and would need to be delivered in hospital.

What about that trebling of the death rate that many news-papers picked up on? If you looked at death rates around the time of birth, there was little difference between babies born at home and in hospital, which suggests it wasn't down to babies failing to be resuscitated or anything like that. The problem crept in if you looked at death rates in the twenty-eight days following birth. The risk was still small: 32 deaths for every 16,500 home births, compared to 32 deaths for every 33,302 hospital births. However, not all of the studies that Wax analysed recorded this 'neonatal death rate', and because the numbers are so small it's difficult to say with certainty just how great the risk is. What's more, when Wax excluded those home births assisted by people other than certified midwives, there was no increase in the death rate between planned home and hospital births. The babies were just as safe, regardless of where they were born.

In 2008, a separate study was published that calculated national death rates among babies born at home in the UK between 1994 and 2003. It concluded that home births were generally safe and not associated with an increased risk of mother or baby dying (if anything, the risk was slightly lower for home births), but that the risk increased if the mother needed to be transferred to hospital due to complications. This study also generated plenty of headlines (some claiming it as proof that home births are safe; others holding it up as evidence that they are dangerous), but its conclusions are difficult to interpret – not least because its definition of women who were transferred to hospital included those who had planned a home birth but then decided to have a hospital birth before labour began (possibly as a result of complications).

Ultimately, then, no one really knows if home births are any more risky than hospital births, because the necessary studies haven't yet been done. This means you should be wary of well-meaning friends, relatives and strangers who tell you that one is more dangerous than the other.

There are several factors you may wish to consider:

• Somewhere between 9 and 37 per cent of women having a home birth will need to be transferred to hospital during labour. The most common reasons include labour progressing too slowly, or the woman needing additional pain relief such as an epidural – and women are more likely to be transferred if it is their first baby. In more serious cases, women may be transferred because of heavy bleeding, concerns about the baby's welfare or because the baby needs emergency care once it is born, and in these circumstances any delay could have serious consequences. You should consider how far it is to your nearest hospital, and how long it would take to be transferred there in an emergency.

- The risks increase if you have a pre-existing medical condition or have experienced complications when previously giving birth, so if your doctor recommends that you give birth in hospital on medical grounds, it is wise to heed their advice.
- Birth is unpredictable, particularly if it's your first time. Try to keep an open mind about your options for pain relief and remember that if you're having a home birth and decide you do need an epidural, there will be an extra wait while you're transferred to hospital.

56

Does walking or squatting speed up labour?

THE TRADITIONAL PICTURE of a labouring woman lying flat on her back, legs in stirrups, screaming in agony is a far cry from how the majority of women spend the early hours of labour. Increasingly, women are encouraged to stay upright for as long as possible, and even to take a long walk, in the hope that gravity will aid the descent of the baby's head and put pressure on the cervix, encouraging it to open.

So strong is this belief that you may find yourself being marched down the corridor by an eager midwife, even if you'd really rather lie down in bed. You may also hear horror stories of how lying down could trigger a downward spiral of medical interventions, starting with you being given drugs to boost the strength of contractions, and ending up with your baby being delivered by forceps or you being wheeled into the operating theatre for an emergency C-section.

Some studies have found that contractions are stronger when women are upright, compared to when they are lying down, so the thinking goes that labour should progress faster if women are upright. Although this is true to some extent, the effect isn't massive. A review of 21 studies which took in a total of 3,706 women found that, on average, the first stage of labour was an hour shorter in women who were upright, compared to those who lay down (the average first stage of labour is twelve to fourteen hours in first-time mums). Women who remained upright were also slightly less likely to request an epidural, but there was no difference in the use of other types of pain relief.

Neither was there much evidence for the idea that lying down would make a natural vaginal birth less likely: rates of un-complicated vaginal births and ones involving mechanical help (in the form of forceps or a ventouse) were similar, and although C-sections were slightly less likely among women who remained upright, the difference was small enough that it could have arisen by chance.

Further analysis found that neither was walking any better than simply sitting or kneeling upright in bed – it was lying down that made the difference. The authors concluded that women should be encouraged to take up whatever position they find most comfortable during labour, while avoiding spending long periods lying down.

Once you enter the second stage of labour and the baby's head has entered the birth canal then squatting may slightly speed things along – although whether your leg muscles are strong enough to maintain this position for long is another matter. A review of twenty studies found that, on average, the second stage of labour (which usually lasts one to two hours) was reduced by about four and a half minutes in women who assumed an upright or squatting position, compared to those who lay down – although the quality of many of those studies was described as poor. There were fewer episiotomies in women who remained upright, although this was offset by an increase in tears to the tissue between the vagina and back passage.

57

Should I push?

ANOTHER COMMON MISCONCEPTION about labour is that it's all about pushing – particularly as the latest evidence suggests that pushing should be reserved only for the final minutes of labour, once the baby's head is already in sight.

Many doctors and midwives believe that pushing as soon as the cervix opens wide enough to let the baby's head through will make birth happen more quickly – particularly if a woman has had an epidural and may be less aware of any natural urge to push. However, that view is being challenged by recent evidence which suggests that it may be better to relax and give the baby time to move gently down the birth canal, with women being asked to push only once the head is visible.

To try to get to the bottom of this issue, an American group recently reviewed seven clinical trials that compared immediate versus delayed pushing in women who had already had an epidural. It found that waiting for up to two hours for the baby to descend before asking women to push decreased the amount of time the women spent pushing and actually increased their chances of having a natural delivery. UK midwives were recently advised to give women with an epidural in place an extra hour to let the baby's head descend, which is an indication that people are starting to take this evidence on board.

Although it has been less well studied, the same principle may also apply to women who haven't had an epidural. A recent trial found that delaying pushing for up to ninety minutes after the cervix had fully dilated roughly halved the total amount of time women spent pushing, without significantly increasing the overall time spent in labour. In other words, you can get there just as quickly with much less effort.

58

Is it better to tear or be cut in terms of healing?

AN IRISH STAND-UP comedian called Dara O'Briain does a sketch about the antenatal classes he attended when his wife was expecting their first child. In it, he describes their teacher's account of how at some point a choice may need to be made between a cut and a tear. 'Obviously,' his teacher tells them, 'you should choose the natural path. Besides which, a tear heals better than a cut.' Recounting this experience to a friend a few days later, the friend says, 'Oh, well, of course, that's well known. Most surgeons these days for the initial incision will use a bear.'

I was also told that it would be better to tear, so I was horrified when after an hour and a half of trying to push my baby out, my midwife turned to me and said, 'I think we may have to make a little cut.' My reaction was to take a deep breath and push with all my might. Out shot my daughter and I was left with a second-degree tear that subsequently became infected and took a long time to heal, so the issue of whether it's better to be cut or to tear is one that haunts me to this day. But the answer is nowhere near as clear-cut as you might imagine.

During the 1970s, episiotomies (a deliberate cut made in the muscular tissue between the vagina and back passage called the perineum) were performed as a matter of routine in order to protect the anal sphincter muscles that control the bowels and to speed up birth. The problem was that this led to many women being cut unnecessarily and suffering discomfort while the wound healed. Also, episiotomies don't always work, because some women who are cut also tear.

These days, the trend is to let nature take its course and only cut a woman if it seems almost certain that she is going to tear, if the baby is in distress or if forceps are going to be used. 'I can honestly say that I have never met a midwife or obstetrician who claims to have done episiotomies when they thought they were unnecessary,' says Philip Steer. 'Unfortunately, there is no straightforward way of determining who is going to suffer an anal sphincter injury other than clinical judgement as to how big the baby is and how tight the perineum is.'

The benefit of an episiotomy is that – particularly if the cut is made at an angle rather than straight down the centre of the perineum – it can direct the injury away from the anal muscles.

In terms of healing, however, O'Briain's antenatal teacher wasn't completely off the mark because, as a general rule, tissue does heal better when it is torn. During a C-section, many surgeons will use a scalpel to cut through the skin and tough connective tissues and to make a small incision in the muscular wall of the uterus, but this is then extended by tearing the tissue.

'This separates the tissues along their natural lines and enables them to come together more easily when the wound is closed,' says Steer.

The problem with tears in the perineal tissue is that the 'natural line' of the tissue often runs straight down the back of the vagina, and – if the tear extends too far – into the back passage. Healing may be less uncomfortable than if an angled cut is made because there is less tension on the wound when you sit down, but there may be an increased risk of serious injury in a minority of cases. Virginia Beckett, consultant obstetrician and gynaecologist at Bradford Teaching Hospital in the UK and a spokesperson for the Royal College of Obstetricians and Gynaecologists, says, 'Personally, I would rather have a nice clean surgical cut that is designed to take pressure away from the important structures down there, than risk an uncontrolled tear, but we don't always get it right.'

59

Does perineal massage work?

IN THE HOPE of preventing a tear, I hovered over the bath each morning during the final weeks of pregnancy, diligently rubbing olive oil into the back wall of my vagina. Since no one had really coached me on what I should be doing, I simply did what seemed to make the most sense: I focused on relaxing my pelvic-floor muscles and stretching the tissue in the way I assumed it would be stretched when my daughter finally put in an appearance. Perineal massage is neither dignified nor relaxing. So is it worth the effort?

The Cochrane Collaboration recently reviewed four studies in which women who massaged their perineum once or twice a week from thirty-five weeks of pregnancy were compared to those who did nothing. Although perineal massage made no difference to the number of tears, fewer of the women who practised it required stitches – at least, among those women giving birth for the first time. It wasn't a massive difference: fifteen women would need to massage themselves regularly in order to prevent one woman from being stitched up, but it's better than nothing. Women were also slightly less likely to require an episiotomy. Massaging once or twice a week seemed to be enough, as massaging more frequently than this didn't improve things.

Unfortunately, perineal massage did nothing to reduce a woman's chances of needing an instrumental delivery (such as one involving forceps or a ventouse) or of developing incontinence, or have any effect on her sexual satisfaction after birth.

If you do decide to try perineal massage, here's how you do it:

- Find a comfortable position, such as lying in bed with your legs bent outwards, standing with one foot on the edge of the bath or sitting on the toilet.
- Using an unscented natural oil such as olive, sunflower or sweet almond oil, insert one or both thumbs into the entrance of the vagina.
- Now, pressing down on the back wall, move your thumbs backwards and forwards in a U-shape, focusing on relaxing your muscles at the same time.
- Aim to massage for around five minutes each time. You can try pushing harder each time you do it.

60

Is there anything else I can do to prevent tearing?

THE SAD FACT is that around 85 per cent of women will experience some degree of tearing during a vaginal birth. This can range from minor damage to the skin around the vagina and labia that heals within a couple of days, to the horror of a fourth-degree tear that rips through all the muscles between the vagina and back passage. Fortunately, fourth-degree tears are relatively rare, affecting just 3 to 4 per cent of women. But tearing is particularly common in first-time mums, if you have a large baby or one that's in a difficult position, or if forceps are used to get the baby out.

Is there any way of preventing or minimizing the damage you might suffer? The Cochrane Collaboration recently reviewed different techniques that midwives sometimes use to reduce the risk of tearing, including massaging the perineum (see 59: 'Does perineal massage work?'), applying a warm compress, placing pressure on the perineum or manipulating the baby's head to slow the rate at which it emerges.

What seemed to be most effective was applying a warm compress to the perineum during the second stage of labour when women are actively pushing. This reduced the risk of serious tearing, helped to relieve pain and reduced the risk of urinary in-continence.

Giving birth in water also seems to reduce the chances that a woman will need an episiotomy, although the risk of a serious tear remains about the same. However, although the idea of tearing during childbirth may make you cringe, recovering from a tear isn't as bad as you might imagine (see 70: 'How do I know if my stitches are OK?').

61

Will delaying cord-clamping benefit my baby?

A PROUD NEW father is handed a pair of scissors and asked if he'd like to cut the umbilical cord. Should he do it, or allow the lifeline that has connected mother and baby for the past nine months to pulsate for just a few minutes more?

Once the baby is safely delivered and starts breathing air, the umbilical cord fast becomes redundant. A series of contractions quickly shuts off the flow of blood to the placenta, which then begins to detach from the uterus wall and follows the baby out into the big, wide world.

In many Western countries, midwives are quick to clamp and cut the umbilical cord. But the placenta contains a large amount of the baby's blood and some have argued that this blood should be given two to three minutes to flow back into the baby's body before the cord is cut. Doing so can increase a newborn's blood volume by around 32 per cent.

Some argue that this extra blood isn't necessarily good for babies: they take longer than adults to dispose of worn-out red blood cells and this can lead to a build-up of a yellow pigment called bilirubin which becomes toxic at high levels and leads to jaundice. However, iron deficiency is also quite common among babies during the first few months of life, and any extra blood they receive at birth should help counter this. Around 3 to 7 per cent of young children in Europe are thought to be deficient in iron, and this can have a negative impact on their brain development.

Several recent studies have bolstered support for waiting a few minutes before clamping and cutting the cord. In one of the largest studies to date, Swedish experts compared four hundred babies whose cords were clamped either ten seconds or three minutes after birth. Four months later, the babies whose cords had been clamped later had higher iron levels and were less likely to be suffering from iron deficiency or anaemia. There was no difference in rates of jaundice between the two groups.

A separate review of eleven studies by the Cochrane Collaboration also found higher levels of iron in babies whose cords were clamped later, but this time there was an increase in the number of those that needed treatment for jaundice (which usually involves placing them under a blanket that bathes the skin in light and helps break down the bilirubin). The authors concluded that so long as access to jaundice treatment is readily available, then delayed cord-clamping should be offered more widely.

Ouch!

62

What's more painful: childbirth or having your leg chopped off?

JUST HOW PAINFUL is labour? It's a question most women ponder at some point during pregnancy, and if you're deliberately trying not to think about it, then I advise you to skip this section.

Personally, I can say that nothing I've experienced compares with the pain of labour – both in intensity and in the type of pain: it's not like the throb of a headache, the shooting pain of standing on a badly sprained ankle, or the raw, searing pain of a bad burn. The experience also seems to vary from woman to woman. Although almost all women will experience lower abdominal pain during contractions, some may also experience continuous lower-back pain or pain in other highly focused areas. It also varies in intensity from woman to woman. The pain occurs for lots of reasons: intense muscle contractions, the stretching and tearing of internal tissues and pressure on the joints of the pelvis.

There's a tendency for antenatal teachers to gloss over the intensity of labour pain, which I think is a mistake, as at least one study has suggested that women whose expectations of labour are violated by the reality of it feel more pain. A recent study of 324 pregnant women found that although 36 per cent anticipated suffering extreme pain during labour, 65 per cent reported experiencing it. If women aren't given accurate information about what they can expect from labour, it is impossible for them to make an informed decision about their preferences for pain relief. On the other hand, fear about pain can also make it worse, and I certainly don't mean to scare people. The reality is that labour hurts, but it's not insurmountable, and there are both coping strategies and drugs that can help you through (see 65: 'Can hypnosis or alternative therapies reduce labour pain?').

So can men ever hope to understand the pain of childbirth? The answer is: maybe. Back in 1981, a pain researcher called Ronald Melzack and his colleagues at McGill University in Canada set out to assess labour pain in 141 women and compare it to other types of pain such as back pain, non-terminal cancer pain or toothache. They used something called the McGill Pain Questionnaire, which consists of a list of seventy-eight adjectives divided into twenty groups that describe different aspects of pain. Patients are asked to circle words to describe how they are feeling: words like 'jumping', 'flashing', 'shooting', 'cutting' or 'lacerating'. Each of the words has been given a numerical score; when these are added together, the result provides a good idea of how much pain someone is in.

Melzack gave the questionnaire to women whose contractions were less than five minutes apart and whose cervixes had already dilated two to three centimetres. More than 33 per cent of the women used the following words to describe their pain: 'sharp', 'cramping', 'aching', 'throbbing', 'stabbing', 'hot', 'shooting' and 'heavy'. 'Intense' and 'tight' were also widely used, and more than 80 per cent of women said their pain was either tiring or exhausting.

A later study by the same group found that 60 per cent of first-time mums described the pain of contractions as being 'unbearable', 'intolerable', 'extremely severe' or 'excruciating'.

The researchers also found widespread variation in overall pain scores between individuals. Women who suffered from premenstrual pain seemed to have a particularly hard time of it – possibly because their bodies overproduce chemicals called prostaglandins which also help to trigger contractions. In addition, first-time mums seemed to suffer more pain than women who had been through labour before, and pain scores were marginally lower for women who had attended antenatal classes which included information about breathing exercises and relaxation. Other studies have also found that first-time mums report more labour pain, particularly during the early stages of labour, when the lower half of the uterus and cervix is being stretched and pulled apart. It's possible that these areas are more supple and transmit fewer pain signals in existing mothers.

So how does the pain of labour compare to other types of pain? When Melzack compared scores from patients battling other conditions or illnesses, labour pain ranked as more painful than non-terminal cancer pain, back pain, toothache or arthritis. But it could be worse: a different study found that having a limb amputated, or a rare condition called causalgia – in which the arms and legs feel as if they are continuously burning and aching – ranked worse than the pain of labour.

63

Do hormones block the pain of labour?

ADVOCATES OF NATURAL childbirth insist that the body produces natural painkillers during labour and to some extent they're correct: contractions do trigger the release of hormones such as beta-endorphin, which binds to and blocks the same receptors in the brain as morphine. The release of beta-endorphin is dampened in women who receive epidural anaesthesia primarily because their bodies are no longer screaming out for pain relief.

Israeli researchers recently found that women's pain tolerance increases during labour, which could be because of chemicals such as beta-endorphin. They used a device called a dolorimeter, which applies steady pressure to tender areas of the body and records the point at which it becomes painful, to record pain thresholds in forty pregnant women before, during and after birth. Women experienced significant increases in their pain threshold once labour was established, and this was particularly true of women whose pregnancies had lasted forty weeks or more. Just why women whose pregnancies reached full term should have greater pain tolerance is unclear, although studies in rats have shown increases in pain tolerance in the days leading up to labour as well as during the actual event itself.

In a separate study, Danish researchers also found that women's pain thresholds increased during the third trimester. In addition, the intensity of labour pain was lower in women who had recently experienced pelvic pain – possibly because the body's natural mechanisms for relieving pain were already activated.

Labour triggers the release of the fight-or-flight hormones adrenaline (epinephrine) and noradrenaline. Often perceived as the bad guys because they can reduce the intensity of contrac-tions, these also reduce pain. One evolutionary explanation for why we release these hormones during labour is that by slowing labour down they might buy women some extra time to seek out a safe and sheltered place to give birth.

Just like beta-endorphin, levels of adrenaline and noradrenaline fall once women are given drugs to relieve pain, but that's not necessarily a bad thing, as women often find that labour progresses more quickly once their pain is under control. However, since panic can also increase the release of adrenaline, it is worth trying other coping techniques such as breathing and relaxation as well.

Finally, other studies have suggested that a lack of family support can heighten women's perception of pain, while women who give birth during the morning seem to experience less pain than those who give birth at night, possibly due to natural variations in the levels of hormones at different times of day. So, if you can manage it, aim to get a good night's rest and plenty of hugs from your partner, and hope for the best.

64

Why do women come back for more?

HOW MANY TIMES have you been told: 'Once you hold your baby in your arms you'll forget all about the pain you have just endured'?

Although it's not true to say that labour pain is immediately forgotten, the memory of it does seem to fade with time – at least in women who have had a positive experience of childbirth. Women who describe childbirth as a negative experience two months after giving birth tend to have fewer subsequent children and take longer to have their next child than those with positive memories.

In a recent study, Swedish researchers asked women to rate their labour pain on a seven-point scale at two months, one year and five years after the birth. Those who had a very positive or positive recollection of childbirth at two months remembered the pain as less intense as the months went by, suggesting that their memories were edited over time, while women who described childbirth as a negative experience didn't change their assessment of labour pain at later dates. Importantly, whether women rated childbirth as positive or negative was unrelated to the amount of pain they had experienced. Instead, it seemed to come down to other factors, such as the level of support they received from the doctors and midwives looking after them.

There's also a physical mechanism by which childbirth might muddy our memories of the experience. Oxytocin, which is released in vast quantities during late pregnancy and childbirth, seems to alter the way in which memories are laid down. Although no one has specifically studied the effect of this hormone on memories of childbirth, other studies in which people have inhaled oxytocin before doing memory tests have suggested that it can reduce their ability to recall lists of words or pictures, while animal studies have hinted that oxytocin can interfere with the consolidation and retrieval of memories. If this is true, it might also suggest that pregnancy-related forgetfulness, or 'mumnesia', has an upside: helping women to forget the pain of labour.

65

Can hypnosis or alternative therapies reduce labour pain?

DESPITE GROWING NUMBERS of C-sections in many Western countries, natural childbirth methods such as massage, birthing pools and hypnosis are increasingly popular. However, because many of these therapies have never been rigorously tested, it can be difficult to know if they are any better than a placebo or nothing at all.

Partly for this reason, the Cochrane Collaboration – an international organization that reviews the results of high-quality clinical trials to help people make evidence-based decisions about their health – recently pulled together all its previous studies assessing different methods of pain relief during labour (reviewing 310 studies in all). These methods ranged from established medical procedures such as epidurals through to injections of sterile water, hypnosis and massage.

After reviewing the evidence, the methods were divided into three categories: those that work, those that may work and those about which there is insufficient evidence to make a judgement. Things that definitely provide pain relief include epidurals and gas and air, but none of the non-drug methods made the grade. Water births, relaxation, acupuncture and massage did make it into the 'what may work' category, however. In most cases, only a single good-quality clinical trial was available to make the judgement, but these methods do seem to relieve pain and improve the experience of childbirth compared to placebo or nothing at all.

In the case of hypnosis, sterile-water injections, aromatherapy, TENS, biofeedback and opioid drugs like pethidine, the Cochrane

Collaboration said there wasn't sufficient evidence to judge whether or not they are effective. That's not to say they don't work, but any studies that exist aren't of good enough quality to prove it.

Let's look at the evidence for some of these alternative methods of pain relief in more detail.

Water birth

Once dismissed as a fad, birth pools are now a common fixture of labour wards in many countries, particularly the UK, where more than three-quarters of hospitals now provide them. Supporters argue that warm water relieves the pain of contractions, reduces the risk of tearing and helps women feel more in control of their birth. However, fears remain about women getting too hot or too cold and of babies being at greater risk of infection or drowning if they are born in water. In the majority of cases, women who use birthing pools get out of the water before giving birth, making some of these concerns irrelevant – but of the studies that have been conducted, there is no convincing evidence of any increased risk of complications among women who give birth in water.

When the Cochrane Collaboration reviewed twelve studies looking at the effects of immersion in warm water, it concluded that, although it made no difference to the rates of C-section, instrumental delivery (involving forceps or a ventouse), the baby's health or the risk of tearing, it did reduce the intensity of pain and women were more likely to feel satisfied with their birth experience.

Transcutaneous Electrical Nerve Stimulation (TENS)

TENS machines comprise a small box that transmits low-intensity electrical impulses to the skin via several electrodes, which are usually attached to the lower back, producing a kind of buzzing sensation. No one really understands why TENS works, but one theory is that the electrical pulses stimulate nerves in the spinal cord that block the transmission of pain. It may also serve as a distraction and give women a sense of control over their contractions.

Plenty of clinical trials have investigated whether or not TENS works for labour pain, but, although a handful of these studies have found some benefit, the majority have found none. For this reason, the UK's National Institute for Health and Clinical Excellence, which issues evidence-based guidelines for doctors, recently stated that TENS shouldn't be recommended for pain relief in labour. The Cochrane Collaboration, which looked at seventeen of these studies, also concluded that TENS had little or no effect on pain intensity or on women's satisfaction with pain relief.

Massage

If you attend antenatal classes, you'll probably be shown some massage techniques that can be helpful during labour. A common one is for the birth partner to use the heel of their hand or their thumbs to firmly massage around the base of the spine, which supposedly counters the pain of contractions. A shoulder or back massage may also help to relax you during labour. When the Cochrane team reviewed six trials looking at whether massage relieves labour pain, they found some reduction in pain during the first stage of labour, but no effect later on.

Acupuncture and acupressure

Acupuncture points commonly used to reduce labour pain are located on the hands, feet and ears – although a scientific explanation for how acupuncture works is still lacking.

The Cochrane Review examined thirteen studies and found that both acupuncture and acupressure reduced the intensity of pain and increased women's satisfaction with pain relief. They were also associated with fewer instrumental deliveries (involving forceps or a ventouse) and fewer C-sections. Although this may sound impressive, just three of the thirteen trials looked at the risk of instrumental deliveries, and just one trial of 120 women recorded the rate of C-sections. Differences between the studies that looked at pain relief also made it difficult to say with any certainty just how effective acupuncture is. Other reviews have similarly concluded that more research is needed.

Hypnosis

Hypno-birthing is a form of self-hypnosis in which you are trained to relax and visualize yourself in another place, such as a beautiful garden, or picture a dial that you can turn up or down to reduce sensation in your body. Women are also taught breathing techniques to calm themselves and aid pushing, and are encouraged to substitute words such as 'pressure' for 'pain', and 'surges' for 'contractions'. The main goal is to reduce tension and fear in the hope that birth will progress more naturally, and, despite being 'hypnotized', women remain fully in control and aware of what is going on around them.

Having used hypno-birthing myself, I can testify that it reduced my anxiety in the run-up to labour, though I'm not sure it was helpful during labour itself. When things got really bad, I was impelled to throw a sponge at the MP3 player that told me for the hundredth time to 'embrace each gentle birthing wave'.

The few studies that have investigated hypnosis during childbirth tend to agree. Although one Cochrane review of five studies found a decreased use of pain-relief drugs and a greater likelihood of a natural vaginal birth among women who used hypnosis, when the Cochrane Collaboration conducted its review of reviews it found no difference in women's satisfaction with pain relief or the childbirth experience compared to women who weren't hypnotized.

Relaxation and yoga

Although somewhat similar to hypnosis, relaxation techniques such as progressive muscle relaxation and yoga do seem to show some benefit. A review of eleven studies of relaxation techniques found that they reduced the intensity of pain and increased women's satisfaction with their pain relief, as well as improving their satisfaction with birth overall.

Aromatherapy

The Cochrane Collaboration found just two good-quality studies looking at the effectiveness of aromatherapy for pain management during labour. These studies, which compared aromatherapy to placebo or no treatment, found no difference in the intensity of pain or the likelihood of having a vaginal birth or a C-section. Neither was there any difference in the use of additional drugs by women in the different groups.

66

Does an epidural make a
C-section more likely?

MOST DOCTORS AND midwives agree that by far the most effective means of reducing pain in labour is epidural anaesthesia. Those who have succumbed to the lumber needle will also testify to the enormous wave of relief that comes with the elimination of hours of exhausting pain. Yet lingering doubts about the safety of epidurals for mothers and babies remain. Some mothers worry it may increase their risk of a C-section or delivery involving forceps or make it more likely that they will tear. Others worry about the drugs reaching the baby and harming it in some way.

Some women also feel that requesting an epidural would be a kind of failure, because it means they couldn't handle the pain. One friend told me that she didn't think she'd be able to look her antenatal classmates in the eye if she had an epidural, never mind the fact that, statistically, many of them were likely to do exactly the same thing. Around 20 per cent of UK women have an epidural, while in the US it is closer to 58 per cent.

Labour is one of the most painful experiences that you're ever likely to go through (see 62: 'What's more painful: childbirth or having your leg chopped off?'), but it also varies a great deal from woman to woman and from pregnancy to pregnancy. Some women undoubtedly do give birth without a great deal of pain, but others need drugs to get through it. If you've never been through labour it is impossible to know how your body will react, so it really is best to keep an open mind about the options available to you.

A recent study of women who made birth plans in advance

of labour found that although more than half of them originally said they'd like to avoid having an epidural, 65 per cent ultimately requested one. What's more, 90 per cent of women who received an epidural reported being pleased with their decision when asked about it later.

An epidural is an injection administered via a tube inserted into the lower spine, close to the nerves that transmit pain signals from the lower body up to the brain. This may sound gruesome, but it isn't as painful as you might imagine (particularly when you're already in the throes of labour), and you're usually given a small injection of local anaesthetic before the tube goes in. Drugs that inhibit those pain signals are then pumped into the spine, providing rapid relief from labour pain. Because the drugs are injected locally, they don't get into the baby's body.

Whereas having an epidural used to mean that you'd be flat on the bed and unable to move your lower body for the rest of labour, modern drug combinations mean that, although you're unlikely to be running around the labour ward, it is possible to get into different birthing positions, such as a hands-and-knees posture, and to bear down and push your baby out during the second stage of labour. Women are also often given a hand-held pump which they can use to top up levels of the anaesthetic drugs and control the amount of sensation they feel.

There are some risks, such as the epidural not working properly, a sudden drop in blood pressure or the accidental puncture of the tough envelope surrounding the spinal cord, which can trigger headaches. Women may also lose the ability to sense when their bladder is full, so a catheter will usually be inserted to empty it (also not as bad as it sounds).

Like so many areas of pregnancy and birth, those studies that have compared the risks of epidural anaesthesia with births without any drugs have reached mixed conclusions, depending on the type of women they studied and the precise question asked. The Cochrane Collaboration recently brought together the results of twenty-three of these studies to try to reach some kind of consensus on the risks and benefits of epidurals. It concluded that they do not increase the risk of C-section, the length of the first stage of labour, long-term backache or have any negative effects on the baby. However, the second stage of labour may last slightly longer (around thirteen minutes, on average), women are marginally more likely to need supplemental oxytocin to boost the strength of contractions, and they are 42 per cent more likely to have an instrumental delivery (one involving forceps or a ventouse).

Because the chances of having an instrumental delivery are pretty low (only around 4 per cent of births involve forceps and 8 per cent a ventouse), this isn't as bad as it sounds. I contacted the authors of the review and asked them to calculate how many women would need to be given an epidural before one of them had an instrumental delivery as a result. The answer is approximately twenty.

Even so, you may feel nervous about any increased risk, because instrumental deliveries tend to be associated with a greater chance of tearing or a cut being required in the wall of the vagina in order to help the baby out. But even though slightly more women who have epidurals have an instrumental delivery, it doesn't necessarily mean that the epidural is to blame. Women who already have an epidural in place are also more likely to be offered mechanical help to get the baby out because their vaginal area is already numb.

A separate study that examined the main risk factors for serious injuries to the anal sphincter muscle that controls the bowels found no association with either epidural anaesthesia or the use of oxytocin to speed up or induce labour.

As for risks to the baby, the Cochrane review found that newborns whose mothers had epidurals in place showed lower signs of distress (as measured by a fall in blood pH called acidosis) – although there was no difference in more severe cases. Neither was there any difference in admissions to the intensive care unit between the babies of mothers who had epidurals or no pain relief, or in their Apgar score (a measure of how healthy the baby looks when it is first delivered) at birth.

67

Do gas and air or pethidine relieve pain?

BESIDES EPIDURALS, the most common drugs offered to women in labour are gas and air (nitrous oxide), and pethidine. Gas and air works not by interfering with the transmission of pain signals to the brain but by making you less bothered by them. It is the most commonly used drug for relieving labour pain and often helps take the edge off contractions, but it can leave you feeling light-headed and sick if you inhale too much of it. The good news is that the effects of the drug quickly wear off and it doesn't affect the baby. A recent review concluded that fewer women who inhaled gas and air reported experiencing severe or extreme pain during the first stage of labour, compared to those inhaling a placebo drug, which suggests that it does work.

The evidence that pethidine provides effective relief from labour pain is more mixed. Pethidine has been available to midwives since the 1950s and is related to the powerful painkiller morphine – although it is certainly less effective than that. It is injected into the muscles and, unlike gas and air or epidural anaesthetic, it can cross the placenta and make babies sleepier, less good at suckling and slower to establish breastfeeding. Large doses can also interfere with the baby's breathing and muscle movement, although it is generally regarded as safe for use in labour and drugs can be given to reverse its effects in the unlikely event of an overdose. Pethidine also tends to make mums feel drowsy and can make you feel sick or dizzy.

When the same review looked at studies of pethidine, it concluded that there was insufficient evidence to judge whether it was more effective than a placebo or other interventions for pain relief in labour. This may seem incredible for a drug that has been around for so long and is used in 39 to 56 per cent of obstetric units in the US, but few high-quality studies have ever been done to establish if it really works. Those that have been conducted have produced mixed results, with some saying it has no effect and others suggesting it does have a modest effect.

The Post-Pregnancy Body

68

What causes the baby blues?

IT'S IRONIC THAT at what could be one of the happiest and most exciting times of a woman's life, she can be sent plummeting into the depths of despair. Yet up to 13 per cent of new mothers experience postnatal depression and 70 per cent experience the more short-lived symptoms of the baby blues, including extreme sadness, mood swings, anxiety, sleeplessness and irritability. I remember being overwhelmed with a sense of not being able to cope and feeling an irrational anger at my husband when he popped out for a couple of hours, a week after the birth of our daughter (even though she was sound asleep). Several friends who had babies at the same time later admitted to feelings of depression in the immediate aftermath of birth – despite putting a brave face on it at the time. Fortunately for all of us, it was a passing phase, but it was only as we prepared for our next round of births that we felt comfortable enough to talk openly about what we went through the first time around.

Only about a quarter of women seek professional help for depressive symptoms after birth. Yet not taking these feelings seriously can have disastrous consequences in terms of women's ability to bond with their child and their own long-term mental health. Children of depressed mothers tend to be more with-drawn, and women who have suffered from postnatal depression are twice as likely to experience another depressive episode within the next five years.

So what's to blame for this often extreme crash in mood? Unsurprisingly, hormones play a big part in it. In the first three to four days after birth, levels of oestrogen drop a hundred- to a thousand-fold, and there are also big falls in progesterone.

Julia Sacher and her colleagues at the Max Planck Institute in Leipzig, Germany, recently discovered that this fall in oestrogen corresponds with a dramatic rise in levels of an enzyme called monoamine oxidase A (MAO-A) in the brain, which breaks down several mood-boosting chemicals, including serotonin, dopamine and noradrenaline. Sacher found that levels of MAO-A peaked five days after birth, which is precisely when many women experi-ence the greatest crash in mood. Although it was a small study, those women with the greatest rise in MAO-A seemed to be at greatest risk of going on to develop postnatal depression. Other studies have found that women who experience severe symptoms of baby blues in the first week after birth are at higher risk of developing longer-term depression. Sacher suggests that eating foods that are rich in the amino acids tryptophan and tyrosine (such as wheat, oats, red meat and seafood) may help reduce the risk, as both are needed to make mood-boosting chemicals such as serotonin. However, if a woman continues to feel depressed or unable to cope much beyond ten days after the birth (when symptoms of the baby blues typically lift), she should speak to a doctor.

69

Why can't I poo?

CONSTIPATION IS A common side effect of giving birth which few women are warned about. It could be two to three days before you do a poo and, especially if you've experienced a tear, have become dehydrated as a result of a long labour or losing blood or didn't eat much fibre in the run-up to birth, that first trip to the toilet can become a mental battleground. Even if you aren't constipated, it may take a while to poo, because pregnancy hormones such as progesterone and the birth itself slow down the movement of digested food through the gut. In addition, the bowels often open during a natural birth (whether or not you are aware of it), which adds to the delay of needing to go.

I had terrible constipation following my first birth and – too embarrassed to say anything to the midwives who came to visit me – I suffered in silence. The result was an anal fissure: a painful tear in the wall of my bottom which, in combination with the vaginal tear I'd experienced giving birth, led to long-term problems that ultimately required surgery to correct. I write this not to scare people but in the hope that others can avoid similar misfortune. If you can't go, you're not alone.

One study of 313 women giving birth for the first time found that twenty-nine (9 per cent) developed an anal fissure, although few of these would have been diagnosed under normal circumstances, as the symptoms are often similar to haemorrhoids and most eventually clear up by themselves (I was particularly unlucky). Less serious consequences of not pooing include abdominal pain and increased discomfort when you eventually do go.

You can avoid constipation by drinking plenty of fluids and eating lots of fresh fruit and vegetables. Once constipation develops, you need to try to go to the toilet as soon as the urge takes you, even if the prospect is terrifying. 'Women are often fearful of opening their bowels after birth because they think it will be painful, so they withhold it and it gets more difficult as a result,' says Virginia Beckett. 'It becomes a self-fulfilling prophecy.'

Besides staying hydrated, Beckett suggests taking a stool softener such as lactulose (which can be bought over the counter in the UK) if you haven't opened your bowels twenty-four to forty-eight hours after the birth.

If you're still struggling to go, experts recommend sitting on the toilet, ideally with your feet on a footstool and your elbows resting on your knees, and relaxing. If nothing happens, you can try to stimulate bowel movements by doing an exercise called bracing and bulging. Put one hand on your lower abdomen and one on your waist and then tighten your abdominal muscles: you should feel your hands being pushed out forwards and sideways (this is a brace). Now do the opposite and push your tummy muscles out to the front (this is a bulge). Repeat ten times, ending with a big bulge. If you're still getting nowhere, don't be embarrassed to speak to a doctor or midwife about your problem.

70

How do I know if my stitches are OK?

AROUND A THIRD of women who have a vaginal birth will experience a tear that needs to be stitched up afterwards – another fact that is often shrouded in silence. Though the prospect of needing stitches in such an intimate area may sound terrifying to the uninitiated, the reality is that, although uncomfortable, the pain of stitches isn't unbearable and in the vast majority of cases women heal up just fine.

A second-degree tear (one involving the skin and muscle behind the vagina but not extending into the muscles around the back passage) will typically take two to three weeks to heal, but the initial pain should have eased after several days. 'A week to ten days after the birth you should be feeling markedly better,' says Beckett. Beckett warns that you may feel a slight 'tightening' sensation around the wound shortly before the stitches dissolve at around ten days as the tissues knit together and start to heal, but this is normal. However, if you start to feel unwell, experience significantly increased pain or heat around the wound or notice a smelly discharge you should speak to a doctor or midwife, as you may have developed an infection. The risk of developing one can be reduced by keeping the wound as clean and dry as possible (which is admittedly difficult, given the location of the injury), and they can be treated using antibiotics. Beckett recommends sitting in a shallow bath of warm water two to three times a day (you don't need to add anything to the water) and patting the area dry afterwards.

Given how common stitches are, little research has been done on the most effective way of relieving the inevitable discomfort they bring. Women are often told to use ice-packs or cool gel pads to reduce pain and swelling, but a recent Cochrane Review concluded that the degree of pain relief was small. Meanwhile, different doctors and midwives may recommend applying witch hazel or adding salt, lavender oil or arnica to the bath, though there's little evidence to suggest any one of these methods is better than another. Paracetamol or ibuprofen can be taken to relieve the pain, and both are considered safe for the baby, as is diclofenac (although this has to be prescribed by a doctor). Doing pelvic-floor exercises and avoiding sitting for long periods can improve circulation to the area and may speed up healing. Probably, the best thing is to try a combination of these different strategies and see what works for you.

Although the initial pain of a tear should have subsided after a week, it is also quite normal to feel some discomfort in the area surrounding the vagina for weeks, or even months, after a natural birth. If this is distressing you or affecting your sex life, you should insist upon speaking to a doctor (see 54: 'Are vaginal births really better than C-sections?').

Even if you don't need stitches, it's usual for the whole vaginal area to feel swollen and tender for a few days after a vaginal birth.

71

Can other countries (e.g. France) teach us anything about getting back into shape after birth?

WE OFTEN ENVY the French. They generally have better weather, better food and are more chic than us. But there's one area where we genuinely should feel green-eyed: postnatal care.

In many countries, women are discharged from hospital and – apart from a few visits from a midwife to check the baby's health – they're largely left to themselves. In France, women see a gynaecologist within the first month of birth, when they're able to discuss any pain or sexual problems they're experiencing. They also routinely get twelve half-hour sessions with a postnatal physiotherapist who specializes in rehabilitating the pelvic-floor and abdominal muscles.

'We focus on bringing the perineum back to health and improving sexual sensation,' says James Turgis, a French physiotherapist who now runs a private physiotherapy clinic called Mummy's Physio in London. The abdominal muscles will also be checked for any damage or separation and women are given exercises to tone them up. It's no wonder French women are back in their pre-pregnancy jeans before they know it (mine are still sadly languishing at the back of the wardrobe).

Although it's normal for the skin of your belly to feel a bit saggy in the months after birth, some women develop 'mummy tummy', a horizontal roll of fat that hangs down in two horns if you bend over. This is often a sign that the abdominal muscles have failed to close back together after pregnancy (the technical name is an abdominal diastasis).

As the baby grows during pregnancy, the muscle that wraps around the front of the belly begins to pull apart. In a few countries, including France and Australia, the extent of the damage is checked before women leave hospital and they are given exercises to correct it. In other countries this simple test is overlooked and women go home oblivious to the fact that there may be a problem.

Besides causing mummy tummy, an untreated diastasis can also lead to lower-back pain and bladder incontinence, and if women start to exercise with a diastasis this can cause further problems. 'Unless the tummy muscles have closed back together then the pelvic-floor muscles don't have a good anchor,' says Maria Elliott, a physiotherapist who runs SimplyWomensHealth in London's Harley Street. 'Many women who were fit before birth consider their stomach muscles are really strong and mistakenly go out running or start exercising and develop problems like stress incontinence.'

Doing sit-ups or abdominal crunches before the tummy gap has closed can cause the muscles to pull even further apart. 'In France, women are initially helped to strengthen pelvic-floor muscles and to close the tummy gap. Once this is achieved, they are able to start abdominal exercises,' says Elliott.

The good news is that you can check your abdominal muscles yourself. To do this, lie flat on your back with your knees bent, put your fingers into your belly button or just above it and press down as you start to roll your head and shoulders up off the floor. You should feel the two sides of the muscle pressing together, and should only really be able to get the tip of one or two fingers into that gap. If the gap is much bigger then you should consult a doctor or physiotherapist. Problems are more common if you've had more than one baby, or if your baby was particularly big.

If you do discover a gap between your muscles, there are some simple exercises that you can use to help close it:

- Lie on your back with your knees bent and, using your hands or a scarf wrapped around your waist, push your abdominal muscles together. Now, keeping your shoulders pressed into the floor, lift your head, exhaling as you do so. Return your head to the floor. Do fifty reps twice a day.
- Still on your back, tilt your pelvis up towards your head and pull your abdominal muscles in while keeping your buttocks relaxed. Then roll your pelvis back down and relax. Repeat ten to fifteen times.
- In the same position, with your abdominals pulled in, add a load by sliding one heel down until your leg is flat on the floor, then slide it back up again. Now do the same on the other side. Repeat ten to fifteen times with each leg.
- Move on to your hands and knees and relax your stomach muscles. As you exhale, pull your belly button up into your body as tight as it can go, drawing your pelvic-floor muscles in at the same time. Hold for about ten seconds, breathing normally, then relax and repeat one more time.

Pregnancy also takes its toll on the pelvic-floor muscles. 'The hormonal changes during the first three months of pregnancy can change the pelvic-floor muscles, never mind whether there has been a C-section or a vaginal delivery,' says Stephanie Prendergast, a physiotherapist at the Pelvic Health and Rehabilitation Center in San Francisco. The muscles may be overstretched and sluggish, or too tight. Conventional pelvic-floor exercises will lift and strengthen the muscles (see 26: 'How do I know if I'm doing my pelvic-floor exercises properly?'), but women with very tight muscles also need to stretch and lengthen them, or such exercises won't have the desired effect.

Detecting tight or stretched muscles is the sort of thing that postnatal physiotherapists specialize in, and it is difficult to know if you have this problem without seeing one. However, Elliott says that all women can benefit from trying to relax the pelvic-floor muscles after birth – particularly if there's any scar tissue, tender points or damage that can cause tension or tightness. To relax these muscles, every couple of days you can massage the whole area from the bikini line right down and around to where the leg meets the buttocks using a light vegetable oil, such as almond oil. The idea is to improve blood flow to the area and promote healing.

If you tore or had an episiotomy, you should also try to gently massage the scar (once it has healed), to desensitize it, hopefully enabling you to return to pain-free sex. The same applies to women who have had a C-section. Elliott says that with the right attention and exercise, most women can return to normal and enjoy a healthy sex life again.

72

Is it normal to go off sex after birth?

UP TO 86 PER CENT of women report sexual problems in the first few months after giving birth, and this is more likely if you have experienced a bad tear. Things should improve with time, but when 481 British women were interviewed six months after birth, 64 per cent said they were still experiencing at least one sexual problem. Vaginal dryness, pain and loss of desire were the most common problems, and most women reported having less sex than before having children, especially less oral sex.

Besides damage to the vagina, breastfeeding can also wreak havoc with your sex life. Many women feel like the breasts are sexual 'no-go' areas during nursing, while the hormone changes that allow you to breastfeed in the first place can make the vagina feel dry and sex uncomfortable. The breastfeeding hormone prolactin decreases sex drive, while at the same time testosterone, which usually boosts sex drive, is suppressed in both men and women after birth.

That's before you consider the extreme tiredness that comes with feeding through the night and the massive emotional changes that come with giving birth and getting used to your post-pregnancy body. In other words, it's perfectly natural not to feel like having sex.

73

Is there anything I can do to stop my breasts sagging after breastfeeding?

SAGGING BREASTS ARE a battle scar of childbearing that most women see as inevitable. Around three-quarters of mums believe their breasts are different after pregnancy, with bigger breasts, loss of firmness and shrinking boobs being the most frequent complaints.

Most blame breastfeeding for these changes. The forum of UK website Mumsnet has at least four threads devoted to the question 'Does breastfeeding make your boobs sag or shrink?' which have together generated more than 347 responses – mostly from women complaining about the effect breastfeeding has had on their bodies. 'Women consulting a plastic surgeon will often attribute the loss of breast shape or volume to breastfeeding, and concerns over changes in breast appearance are consistently ranked among the most important reasons women elect not to breastfeed their infants,' says Brian Rinker, professor of plastic surgery at the University of Kentucky in Lexington, US.

Surprisingly, breastfeeding isn't the true culprit at all. At least two studies of women describing breast changes after pregnancy have found that breastfeeding wasn't a significant factor, as women who didn't breastfeed also described their breasts as different. 'We found that pregnancy was strongly associated with sagging, and this effect got worse with more pregnancies,' says Rinker, who led one of the studies. 'Being heavier, having larger breasts and cigarette smoking were also associated with more sagging, but a history of breastfeeding, curiously, was not.'

So how does pregnancy cause the breasts to sag? Regardless of whether you eventually breastfeed, the breasts go through big structural changes during pregnancy that can stretch the skin and ligaments. Even as early as six to eight weeks you may notice your cup size beginning to increase and, as pregnancy progresses, the network of lobes and ducts that will eventually produce milk begins to grow and mature. By the end of the second trimester the breasts are already up to the task of producing milk, but high levels of the hormone progesterone inhibit it. Progesterone falls once the baby is born and levels of the breastfeeding hormone prolactin increase, ramping up milk production.

When women stop breastfeeding, these physical changes begin to reverse themselves through a process called involution. First, milk-producing cells begin to self-combust, then the connective tissue surrounding these cells gets remodelled and replaced with fat. This remodelling continues as women age and progress through the menopause, when some notice that their breasts shrink in size.

Just how quickly this all happens varies from woman to woman, and some find that their breasts end up smaller after pregnancy than when they began. Precisely why this happens to some women and not others is unclear, but experts suspect it may come down to how easily you lay down fat on your breasts, which may compensate for lost breast tissue.

Although it's possible that wearing a supportive bra during pregnancy may help guard against future breast-sag or tissue loss, Rinker is sceptical: 'I think the major factors are genetics, age and pregnancy. I don't think exercise or consistent bra use can really help.'

74

Do women who breastfeed really lose their baby-weight faster than those who bottle-feed?

MOST WOMEN LAY down around 4kg of fat during pregnancy – usually on their hips and thighs – which is thought to serve as an energy store for breastfeeding once the baby is born. Breastfeeding is frequently touted as a means of shedding this baby-weight once pregnancy is over, as exclusive breastfeeding eats up around 595 calories per day during the first two months, and 670 calories after that.

The trouble is, we're often told to eat extra calories to make sure our milk supplies don't dry up, never mind the fact that breastfeeding can make you ravenously hungry (and it's not just greed: the breastfeeding hormone prolactin boosts appetite).

So does breastfeeding really help you shed those excess pounds? Although a number of studies have looked into this issue, their conclusions have been less certain than the pro-breastfeeding lobby might have us believe. One problem is that many studies have used mums' own estimates of their weight before and after pregnancy, rather than measuring it directly. Even in studies that have found a link between exclusive breastfeeding and weight loss, the effect is pretty small. A review of five studies that directly weighed and measured women at various points after birth found that after twelve months the difference in weight loss between women who breastfed the longest and those that didn't breastfeed at all ranged from 0.6kg–2kg (a quarter of a stone at best).

If you do decide to breastfeed and want to speed up your weight loss, the good news is that this can be done without compromising your milk supply. One study found that limiting calorie intake to around 2,092 calories a day (which is roughly what a healthy woman should have been eating before pregnancy if she wasn't dieting), combined with around forty-five minutes of aerobic exercise four times a week, should enable women to lose roughly 0.5kg per week without affecting their milk supply – although it would be sensible to wait for your body to recover and for breastfeeding to become properly established before attempting to do this.

75

Do cells from my baby live on in my body after birth?

HAVING A BABY changes you in more ways than you might imagine. From early pregnancy, you will forever carry a little piece of your baby around with you in the form of foetal cells that continue to circulate in your blood and survive in your bone marrow. Precisely what these cells are doing remains unclear, but it increasingly looks like they may help to protect us against diseases, including cancer: a kind of foetal insurance policy that boosts the chances of Mum surviving long enough to support her offspring into adulthood.

This transfer of foetal cells starts from around four to six weeks of pregnancy (before many women even realize they're pregnant) and continues until birth. Even though these cells are foreign and should trigger an immune reaction in the mother, they don't appear to – although some have suggested that as well as protecting Mum against disease, they may occasionally contribute to auto-immune diseases such as rheumatoid arthritis or lupus.

Foetal cells have been recovered from the livers of women with hepatitis C, suggesting that these cells may have lodged there and transformed into liver cells to help patch up the damage caused by the disease. Meanwhile, women with breast and other cancers seem to have fewer foetal cells in their blood than healthy women, hinting that this foetal repair kit may help protect women against the disease – possibly by helping the immune system to identify and destroy cancer cells before they take hold. In addition, animal studies have shown that foetal cells can travel into the brain and mature into different cell types, including neurons (which transmit electrical signals throughout the brain) and the cells that nurture and support them.

It works in reverse, too. Babies get a transfusion of their mother's cells during pregnancy, and these can also stick around for years. There's even evidence that these maternal cells may help baby girls to have a successful pregnancy of their own once they grow up. Pre-eclampsia is a serious complication of pregnancy affecting 6 to 8 per cent of women and is characterized by high blood pressure and protein in the urine. Its exact cause is unknown, but one theory is that it is triggered by the mother's immune system over-responding to the pregnancy. Hilary Gammill and her colleagues at the Fred Hutchinson Cancer Research Center in Seattle, US, recently investigated how a woman's risk of pre-eclampsia related to the number of maternal cells she harboured.

In general, the number of these maternal cells increased as pregnancy progressed, suggesting that an existing population of maternal cells was being either expanded or mobilized from some-where in the body. But whereas around 30 per cent of healthy women had detectable levels of maternal cells in their blood by the third trimester, no such cells could be found in the women with pre-eclampsia, hinting that the maternal cells may be having some kind of protective effect. One possibility is that they may help train a woman's immune system to be more tolerant of cells from the foetus. In other words, Grandma may be indirectly helping to safeguard the survival of her family long after her own body has ceased to bear children.

Babies

Portrait of a Newborn

76

Does birth distress the baby?

It was Sigmund Freud who first hypothesized that all anxiety could be traced back to the original trauma of being born. You could certainly imagine that birth would be a distressing event. You go from spending nine months floating in a warm cocoon to suddenly being rocked and squeezed by muscular contractions that grow increasingly violent until you find yourself being pushed down a narrow tube with such force that the bones of your skull begin to overlap. Scientists have calculated that birth exerts the equivalent of 2kg of pressure on every centimetre of the head, which sounds pretty painful.

Since none of us remembers being born and newborn babies can't speak, it's impossible to know for certain what newborns think of the experience, but they seem to tolerate it relatively well. Recent studies have shown that newborns have high levels of stress hormones called catecholamines in their blood, which may exert a sedative-like effect, as well as high levels of the natural painkiller beta-endorphin, which blocks pain signals in the brain.

Newborns that have endured a vaginal birth also seem to be less sensitive to pain than babies born by C-section – at least until the effects of these natural drugs wear off. Many newborns are given a routine injection of vitamin K shortly after birth, which usually prompts them to screw up their faces and scream. But when Swedish researchers compared how babies born vaginally and by C-section reacted to these injections, and to a cold spoon being pressed against their bellies, they recorded less extreme reactions in those that had been through a vaginal birth. The heart rate of babies delivered vaginally also remained relatively stable compared to those of babies born by C-section, which shot up in response to the pain.

However it's not only endorphins that might blunt pain during birth. Oxytocin is also released in large quantities during labour, and some of it will find its way across the placenta and into the baby. French researchers recently tested how newborn and two-day-old rat pups responded to painful stimuli and found that the newborns had far higher pain thresholds than the older rat pups. When newborn rats were given drugs that blocked the action of oxytocin, they became more responsive to pain.

Still other studies suggest that oxytocin temporarily dampens a baby's brain activity immediately before birth by inhibiting the ability of nerve cells to communicate with each other. As well as reducing the transmission of pain signals, this would have the added advantage of reducing the brain's energy and oxygen needs at precisely the same time as its supply of oxygen is most likely to be disrupted: as the baby is squeezed through the birth canal.

All this should provide comfort for parents whose baby needs additional help to be delivered, be it through the use of forceps that squeeze the head even further, or through a vacuum being applied to the top of the head that pulls the baby out. The chances are they're dosed up on natural painkillers, so it won't be too traumatic.

77

Why do newborns look like their dads?

IT'S VERY UNFAIR. Women spend nine months being pregnant, have the difficult job of pushing the baby out and then feeding it, only to be told constantly: 'Ooooh! She looks just like her dad.'

The jury is still out on whether the resemblance between a baby and its father is real or imaginary. One famous paper published in *Nature* in 1995 found there was a link. Nicholas Christenfeld and Emily Hill of the University of California in San Diego gathered photos of twenty-four people when they were one, ten and twenty years of age. They then showed the photos to 122 volunteers and asked them to match the same children at different ages, which they were very good at. But when photos of three possible parents were introduced, things became interesting. Generally, the volunteers found it difficult to match children to either parent, except for when it came to matching pictures of one-year-olds to their dads (they couldn't match one-year-olds to their mums, though). Here, they guessed right 50 per cent of the time, compared to the 33 per cent one would expect if they were taking a random guess.

This is what evolutionary psychology would predict: babies that look more like their dads are less likely to be rejected by them and are therefore more likely to survive. Given that up to 10 per cent of fathers are thought to be cuckolds, it would be a neat trick.

However, disgruntled mums may take comfort from several later studies that contradict this. In one, Robert French of INSERM in France and Serge Brédart of the University of Liège in Belgium tried to replicate the original study using photos of one-, three- and five-year-old children and found that people were able to guess the mothers and fathers equally well.

78

What do newborns know?

OUT POPS A newborn baby, bleary-eyed and baffled as it tries to make sense of its new surroundings. Most people assume that newborn babies are blank slates born with no knowledge of the wider world. What could they know? They've never seen any of this before.

But babies seem to come with a certain wisdom built in. Some of it may be genetically hard-wired, while some of it may have been accrued through the baby's time in the warm, dimly lit uterus, where it has been able to see, hear, feel and taste for several months.

Some people may wonder how you go about testing what a young baby knows and feels, as they obviously can't tell us with words. There are a few techniques that researchers use, which, although less than perfect, may give us some indication of what's going on inside those tiny heads. One is tracking the movement of a baby's eyes or head, as newborns tend to be attracted to interesting sights and sounds – particularly if they haven't seen something before. They will also turn their head to fixate on something they like, such as their mum's face. Another way of tuning in to a baby's preference is by using a device that allows them to choose between two different stimuli (such as their mother's and a stranger's voice) through the intensity of their sucking. Babies can control the rate and pressure at which they suck, and do it even when they're not hungry, so they can be taught to turn on a certain stimulus if they suck hard enough.

Researchers can also measure a baby's emotions by looking at its facial expressions or the intensity of its cries, particularly if these are combined with physiological measurements such as recording its heart rate or the release of stress hormones such as cortisol and adrenaline.

Here's a guide to the seven pillars of innate baby wisdom. Newborn babies already understand:

Who their mother is

Babies seem to recognize their mother's voice, and possibly her smell, from birth, having already spent months becoming accustomed to them both. Babies begin to hear during the second trimester and, although the sounds of the outside world will be distorted by the time they reach their tender ears, many do get through. Particularly prominent will be Mum's voice, as its vibrations are transmitted through her bones to the uterus.

In the 1980s, a psychologist called Anthony James DeCasper showed that babies only a couple of hours old already knew the sound of their mother's voice. He tested this by creating a feeding contraption that enabled babies to hear a recording of their mother's voice reading a Dr Seuss story called *To Think That I Heard It On Mulberry Street* or a stranger's, according to how hard they sucked. The newborns had a stronger preference for their mother's voice, despite being just a few hours old.

Using the same contraption, DeCasper also showed that babies who had been read a Dr Seuss story called *The Cat in the Hat* during pregnancy preferred to listen to this over an unfamiliar story. Although they probably don't recognize the words, it's quite possible they tune into the rhythms and intonations.

Newborns also seem to be attracted to their mother's smell. In a separate study, thirty newborn babies were each placed on their mother's chest within minutes of birth, but, before this happened, one of the mother's breasts was washed with unscented soap in order to remove its natural odour. Twenty-two of the babies showed a clear preference for suckling at the unwashed breast rather than at the clean one. Another study found that five-day-old babies presented with either a clean cotton pad or one that had been worn next to their mother's breast turned to and started wriggling towards the pad that smelled of their mother (see 83: 'Does skin-to-skin contact really soothe my baby?').

Knowing Mum's voice and smell must be a great source of comfort to a newborn thrust into an otherwise unfamiliar world, and within a couple of hours they recognize their mother's face as well. Although babies seem to be born with an innate appreciation of the human face (see page 173), several studies have found that they rapidly start to discriminate between different faces. In a study conducted at Glasgow Hospital in Scotland during the late 1980s, two new mums with similar hair and skin colour were sat next to each other behind a clear plastic screen, to mask their smell. Their babies were then brought into the room one by one and held 30cm from the women's faces. An independent observer watched as the babies turned their heads from one woman to the other, before eventually fixating upon one of them. Despite their poor eyesight and being only between twelve hours and four days old, the babies knew who their mother was and turned to gaze at her with adoration.

What it is to be human

For at least a few weeks before birth, a baby will have had the opportunity to see its surroundings – particularly on a hot summer's day, when its mum may have exposed her heavily pregnant belly to the sun, allowing more light to penetrate through the layers of skin, fat and muscle and into the uterus.

What a baby hasn't seen is a human face, yet from the moment they are born newborns seem hard-wired to know that a triangular pattern of blobs roughly corresponding to the position of the eyes and mouth has special significance. Studies have consistently shown that newborn babies will turn their heads to look at face-shaped patterns over random patterns, and normal faces over faces where the position of the features has been distorted. They'll also spend longer looking at them. Even nine-minute-old babies are more likely to turn their heads and follow a face-like pattern over a pattern containing the same features scrambled into a different arrangement. It's as if they're born with an inbuilt face-detector.

They also seem to understand that open eyes have some special significance. At Addenbrookes Hospital in Cambridge, UK, researchers took 105 babies who were only a couple of hours old and showed them two pictures: one of a woman with her eyes closed, the other of the same woman with her eyes open. The babies spent significantly longer looking at the picture of the woman with open eyes.

Stranger still, newborns seem capable of relating their own facial features to those of another person. In a separate study, Andrew Meltzoff of the University of Washington in Seattle filmed newborns as an adult opened his mouth or stuck his tongue out at them, then a third person (who couldn't see the faces the adult was pulling) rated the expressions the babies made. Time after time, these tiny babies who were just a few hours old appeared to mimic the facial expressions of the adult. This is no mean feat, considering the babies needed not only to recognize that the object in front of them was a face and that the moist pink appendage sticking out at them related to their tongue; they also required the coordination to stick out their own tongue in response.

Within a few days, newborns also seem capable of discriminating between different types of facial expression. If you show babies pictures of happy or fearful faces, they will spend longer looking at happy faces – although this could be because the predominant expression of the faces they have seen during their first few days are the adoring and happy faces of their parents, rather than because of some kind of innate desire to look on the brighter side of life.

Numbers

It will be some time before your baby can count to ten, but babies are born with an appreciation of numbers as abstract concepts nevertheless. This means they understand that seven objects are different to two objects, regardless of what those objects are. Until recently, people had assumed that a concept of numbers was something you learn, not something you're born with.

One way of testing this is to see if babies can make the link between seven objects that they see and seven sounds that they can hear. Véronique Izard and her colleagues at Harvard University took sixteen babies aged seven hours to a hundred hours and played them a tape of spoken syllables repeated a set number of times, for example, 'ba ba ba ba ba', while showing them collections of circles or squares on a screen. Sometimes the number of objects tallied with the number of sounds, other times it didn't, but the babies seemed to spot when they matched and spent significantly longer looking at the screen.

How do they do it? Other recent studies in both animals and humans have suggested that the brain contains specialized cells called accumulator neurons that fire in response to collections of objects – so, the more things you see, the more cells fire.

Movement and distance

Imagine being cast into the big wide world for the first time. Suddenly you go from seeing very little to being bombarded with lights, sights and movement. But newborns come equipped with some built-in ideas about what they should pay attention to if they're to make sense of the world. For one thing, they know that movement is important. Move a toy back and forth in front of a newborn's face and they will follow it with their eyes. Studies that have tracked the movement of newborns' eyes have also found that newborns pay more attention to the edges of objects, rather than their interiors.

However, babies also know that if objects are passing their eyes as if they are moving forwards, then their bodies should be moving forwards too. Have you ever had that sensation when you're sitting on a train and the train next to you begins to move? For a moment, you feel as if your train is the one that's moving – and if you're standing upright you may lurch forward yourself. Something similar happens if you sit a newborn in a room where the walls can move independently of the floor. Even though the baby is sitting still, if the walls begin to move towards them, its head instinctively moves backwards.

Similarly (and possibly because of this), newborns also understand that objects should get bigger as they move closer to them – something called size constancy. Imagine you're shown a cube, then you're shown the same cube, only twice as far away. You'd probably realize it was the same cube, even though it was only half the size of the original one. So do newborn babies. Alan Slater and his colleagues at the University of Exeter, UK, took two-day-old babies and showed them variations of either a small or large cube at different distances from their eyes. The babies spent longer looking at the odd cube than the cubes that fitted the pattern, suggesting that they appreciated there was something wrong.

The fundamentals of language

Although babies don't speak their first meaningful words until around the age of one, a newborn already has some appreciation of what makes his or her native language unique. Just days after birth, babies can tell the difference between different languages, even if the two languages are spoken by the same person.

They also seem to be particularly tuned to tones of voice conveying happy emotions. In one study, newborn babies were played recordings of the same passage of script read in a happy, sad, angry or neutral voice. The newborns were more likely to open their eyes and look as if they were paying attention when they heard the happy voice, compared to when the same passage was read in the other tones of voice. However, this happened only when the speaker used the baby's native language. If the child of an English-speaking mother heard a happy-sounding

Spanish woman, they paid little attention to her. One possible explanation is that foetuses can detect physical signs related to different emotional states – such as anger prompting an increase in breathing, heart rate and muscle tension – and they learn to associate these states with particular sounds.

Newborns also seem to cry in a melody befitting their native tongue. A recent study of sixty French and German newborns found differences in the intonation patterns of their cries. Whereas the wails of French babies start low and end on a high note, German babies cry in a falling melody – patterns that chime with the intonation patterns of adult speakers of these languages. Since crying may be the only sound a newborn baby knows how to make, this may sometimes represent their first attempt at communication, the researchers suggest. At any rate, it suggests they have been paying attention to language during pregnancy.

Empathy (or the difference between themselves and others)

OK, it'll be several years before your child truly grasps the notion that other people have thoughts and feelings of their own (see 127: 'When do babies develop a sense of themselves as individuals?'), but even newborns react to the emotions of others. Have you ever noticed how, if one baby starts crying, it sets every other baby in the room off too? They're not just responding to loud noise.

Back in the 1970s, researchers found that if they played a newborn a tape-recording of other babies crying, then it started crying as well. When they did the same thing with recordings of equally loud and intense non-human sounds, the babies cried less. So babies seem to know they're babies, and get upset when others are upset.

In a later study, Italian researchers played newborn babies recordings of their own cries and the cries of other babies to see how they responded. The babies showed more visible signs of distress, such as knitting their eyebrows together, screwing up their eyes and grimacing, when they heard the other babies cry, compared to when they heard their own cries, suggesting that they could discriminate between the two. Some people have interpreted this as evidence that the early roots of empathy are present at birth, while others believe it may just be a primitive response to a distress call – similar to the way birds and other animals signal to each other when a predator is approaching.

By the age of eighteen months, most babies have begun to grasp the concept that other people have different likes and dislikes to them, and they will react accordingly. For example, if you show a baby a bowl of crackers and a bowl of broccoli, most infants will make a beeline for the crackers, but if you make a face to show that you love broccoli and pull a disgusted face at the crackers, an eighteen-month-old will generally offer you some broccoli.

However, true empathy, when you notice someone else's distress and are moved to try to do something to make it better, doesn't develop until around the age of two. In the book *How Babies Think*, developmental psychologist Alison Gopnik describes her son's reaction one day when she came back from a particularly tough day at the lab to find the chicken legs that she'd planned to cook for dinner still frozen. Like any frazzled mum, she broke down in tears on the sofa. Upon seeing this, her son, who was approaching the age of two, looked thoughtful and then ran to the bathroom and came back with a large box of plasters, which he proceeded to stick all over her to try to make her better. I have a similar story about my daughter, Matilda, when she was twenty-two months old. She had been to the doctor's, and the next day, I returned home from work heavily pregnant, utterly exhausted and with pains in my belly, and slumped on the sofa. Matilda climbed on to my lap and patted the plaster on her arm. With a genuine expression of concern etched across her little face, she patted her arm and told me: 'Mummy need medicine from doctor.'

A natural sense of rhythm/music

If you're going to be a dancer or a musician, you need to sense the beat. And a baby's ears seem to be tuned to detect rhythm from the moment it is born.

Since babies can't exactly click their fingers in time with a tune, István Winkler and his colleagues at the Hungarian Academy of Sciences in Budapest monitored their brain activity instead. If an adult is played a regular rhythm and that rhythm is disrupted, then their brain produces a characteristic signature which can be detected with an EEG cap that uses tiny sensors to pick up electrical signals.

To see if newborn babies showed the same response, Winkler played them variations of a rock rhythm involving a snare drum, bass drum and a high-hat cymbal. When beats that were un-important to the overall rhythm were omitted the babies showed no response, but when the critical down-beat was absent, the babies' brains detected this disruption to the rhythm.

Researchers had previously suggested that babies learn rhythm as a result of being rocked by their parents, but Winkler's study suggests that a sense of rhythm may be innate. Another possibility is that the foetus learns about rhythm while in the uterus by listening to its mother's heartbeat and the sounds that filter in from the outside world.

An appreciation of rhythm isn't just important for music. It may also help babies to learn about language by disentangling the speech patterns that convey certain emotions or set some languages apart from others. Similar studies looking at babies' brain activity have suggested that newborns can detect differences in pitch and whether a series of notes is rising or falling in scale.

79

How much can newborns see?

OPEN MOST BABY books and you'll read that newborns can only focus on objects 30cm in front of them, which, coincidentally, is often the distance they're held away from the mother's face. This is a myth born out of a scientific paper published in the 1960s, the conclusions of which have since been overturned several times.

Actually, newborns can focus on objects at any distance, but they're not very good at it because the brain has not yet fine-tuned the connections that allow it to adjust the thickness of the lens and bring things into sharp focus. It's a bit like letting a very clumsy person loose with a manual-focus camera – they'll often over- or under-shoot, and the image will come out blurred.

Very young babies sometimes give the impression that they are focusing on nearby objects because they also struggle to co-ordinate the movements of the left and right eyes and keep them pointing in the same direction, which means they sometimes look slightly cross-eyed. Their attention also tends to be drawn to objects close at hand, such as Mum's eyes, mouth, nipples and their own hands. 'Because they may not yet have good control of eye coordination, when they try to focus on close objects they can over-converge their eyes, making it appear to us that they are focused very, very close indeed,' adds Russell Hamer, an affiliate scientist at the Smith-Kettlewell Eye Research Institute in San Francisco. Newborns also struggle to integrate the input from both eyes to produce a 3D picture, so, for the first few months, they see the world in flat 2D.

Another reason why even nearby objects will appear grainy to a newborn baby is that other visual areas are not yet fully developed. These include the retina – the area at the back of the eye on to which images are focused – and regions of the brain that interpret these images. For example, an area in the middle of the retina called the fovea is particularly important for seeing details, but it takes many months before enough receptor cells (called cones) have developed and are mature enough to signal very fine details to the brain.

When Hamer and his colleagues measured newborns' visual acuity, or their ability to see details, they discovered that it is around six times poorer than that of adults. To put it another way, if you think of those eye charts that are often used by opticians to test your eyesight, the only letter that newborns would be able to make out clearly would be the 'E' on the top line. By four months of age, this aspect of their eyesight is much better, but it's not until they reach eight to twelve months that babies' eyesight approaches adult standards. By three years or so, they finally achieve 20/20 vision.

That's not to say, however, that babies can't see an awful lot. Even if it can only make out that 'E' on the eye chart, a newborn baby can see two flies sitting on a white wall from about 2.3 metres, Dad's smile from about 9 metres, or a 70-metre Boeing 747 from more than ten miles.

'A young infant's acuity is good enough to see most of the

things that are important to him or her,' says Hamer. 'Even a newborn infant held in your arms can easily see your eyes, your lips and smile, your nose. They can also see their own hands, fingers, feet and toes.'

There are other differences between how babies see the world and how most adults do. The fovea is also important for colour vision, so as well as the world appearing blurry, newborn babies can't properly perceive colour until they're around four months old – although one study suggests that they are able to tell the difference between red and green from as young as two weeks. Difficulties may start to creep in once they try to distinguish between subtle shades such as light blue and turquoise, however.

New parents are often told they need to buy toys with large, high-contrast black-and-white patterns so their baby can see them and to stimulate visual development, but modern research suggests that this isn't correct. This was based on the belief that babies had little or no colour vision (which isn't true) and that their sensitivity to low contrasts was very limited. However, studies have shown that, for reasonably large patterns, the visual areas of even very young infants' brains can respond to low contrasts, and by nine weeks, babies are only two times less sensitive to contrast than adults. 'By about two months of age your baby is capable of perceiving almost all of the subtle shadings that make our visual world so rich, textured and interesting: shadings in clouds, shadows that are unique to your face; they can even see a white teddy bear on a white couch,' says Hamer.

80

Why do newborns have blue eyes?

EYE COLOUR IS inherited from your parents and is determined by the amount of a pigment called melanin that you produce. People who don't produce much melanin have blue or grey eyes, while those who produce lots of melanin have brown eyes.

However, at birth, very little melanin is being produced – regardless of the genes you have inherited. This is why most babies' eyes appear blue or blue-grey. As babies' eyes are exposed to light, production of melanin is ramped up, so if babies are genetically programmed to produce lots of melanin their eye colour will begin to change as the months go by. A baby's eye colour usually settles down by around the age of six months and is unlikely to change much after this.

81

How do babies go from breathing nothing to breathing air?

A BABY'S BODY undergoes several massive changes in the minutes and hours after birth. Before birth, the baby has 'breathed' by exchanging oxygen and carbon dioxide with its mother's blood via the placenta. Now it needs to switch to breathing through its lungs, but in the womb these are deflated and filled with amniotic fluid – hardly ideal for breathing air. Healthy babies have 80–100ml of fluid in their lungs which they need to get rid of in order to breathe. Around a third of this is squeezed out of the lungs as a result of contractions and being pushed through the birth canal during labour, which compresses the chest. Labour also triggers the production of chemicals called catecholamines, which tell the lungs to start absorbing fluid. Finally, as the baby's chest emerges from its mother's body, it naturally expands, drawing in air. This displaces the remaining fluid from the lungs – much as blowing through a straw into a full glass of water will cause some of the water to be pushed out of the glass and on to the table.

But what about babies delivered by planned C-section? Their mothers don't go through labour, so how does the baby know to start breathing? Part of the answer lies in the clamping of the umbilical cord, which stops the flow of oxygen and triggers a reflex that causes the baby to gasp for air, inflating the lungs and displacing some of the fluid. But most babies delivered by C-section will start breathing before the cord is clamped, so there must be other mechanisms at work. The physical act of removing the baby from the uterus causes the uterus to contract, and this squeezes the placenta and reduces the flow of the baby's blood into it. These contractions also trigger the placenta to start peeling away from the wall of the uterus (as happens in a natural birth). Studies in animals have suggested that exposing the nose to oxygen in the air is enough to initiate breathing, while being brought from the warmth of the mother's body into a cold operating theatre also seems to stimulate this reflex.

For about four months before birth, the baby's lungs have been manufacturing a substance called surfactant, which helps repel water from the lining of the lungs and enables them to inflate and take up oxygen. Nevertheless, babies born by C-section may still have some fluid on the lungs, so they may be watched for any signs that this is causing breathing difficulties.

As the lungs fill with air, this activates another big change in the baby. Before birth, very little blood is pumped from the heart to the lungs, whereas in adults, the entire right side of the heart is devoted to this purpose – sending the deoxygenated blood it receives to the lungs to be loaded up with oxygen, which then re-enters the left side of the heart to be pumped back out to the rest of the body. Before birth, there is a hole between the right and left sides of the heart which largely cuts out the lungs from the circulation. But as the lungs fill with air, this fundamentally changes their structure and makes it far easier to pump blood through them. Because of this, the blood pressure in some of the vessels feeding the heart changes and this prompts a one-way flap of tissue to slam shut over the hole that previously allowed blood to pass from the right to the left side of the heart. The baby's heart and lungs are now plumbed in to work like an adult's.

The tiny baby breathes a sigh of relief. Its transition to life outside the womb is complete.

82

Why do newborn babies smell so good?

THERE'S SOMETHING ABOUT that milky newborn smell that intoxicates parents. Many admit to spending hours inhaling their baby's odour, while, shortly after the birth of her twins, Emme and Max, Jennifer Lopez allegedly told her fragrance experts that she'd like her new perfume, Live, to smell like the top of a baby's head.

It's not just in our imagination. Several studies have shown that women can pick out the scent of their baby from other babies within just a few days of birth, while other research has shown that new mums are more attracted to the smell of newborn infants than are non-mums.

In one study, Alison Fleming and her colleagues at the University of Toronto in Canada asked women to rate the smell of unfamiliar babies' T-shirts on a scale of attractiveness. They found that new mums rated the smells as more delicious than did non-mothers or the mothers of slightly older babies – so there may be something about being a new mum that makes you more tuned into that new-baby smell. Indeed, in a later study, Fleming's team found that mums with higher levels of cortisol – a hormone that has been linked to other aspects of maternal instinct (see 31: 'Are some women naturally more maternal than others?') – were more drawn to baby odours and better at recognizing their own baby's smell than women with lower levels of cortisol.

'We also tested fathers, and they, too, like the odours of babies,' says Fleming. What's more, parents' attraction to these baby body odours seems to increase the more time they spend with their child, and their ability to recognize their own baby's smell also seems to improve with experience.

Although we don't know precisely where these smelly signals are coming from, we can hazard a guess. Small oil-producing glands – sebaceous glands – are found all over the body, but especially on the head and face, and these tend to be particularly active in newborn babies. Sweat-producing eccrine glands are also present in high numbers on the heads of newborns, and secretions from both these glands serve as nutrients for bacteria – whose action is a key factor in the production of body odours, says Richard Porter, an expert in human smell at the French National Centre for Scientific Research in Nouzilly. In Fleming's study, the clean T-shirts were wiped over the babies' heads and bodies before they were given to the mums to sniff.

Body odour can also be influenced by what you eat, so breast-feeding (which is known to affect how babies' nappies smell) may contribute to that delicious new-baby smell. This may also explain why babies stop smelling so unique once they start eating solid food at around the age of six months.

What's more, body odour also has a strong inherited component. 'Mothers may also be able to detect a similarity between the baby's odour and the smell of other family members such as the infant's father or older siblings,' says Porter. Other studies have shown that people can match parents to their children on the basis of their smell, while sniffer dogs have difficulty discriminating between identical twins who share the same genes.

83

Does skin-to-skin contact really soothe my baby?

WHAT NEW MUM doesn't dream of the moment her baby will be placed, sticky and warm, on her naked chest? But does the baby know or care what happens to it during those first crucial moments in the outside world? Although it may sound like a fluffy, New Age, feel-good kind of thing, skin-to-skin contact has been surprisingly well studied by scientists, and the general consensus is that it really does make a difference to the baby's well-being.

Place a newborn baby on its mother's bare chest, and a few interesting things will start to happen. First, the baby begins to make sucking and rooting gestures with its mouth, then its arms will make crawling movements and its legs will start to push against the skin as it slowly inches its way towards the breasts.

A recent review of thirty studies looking at the effect of skin-to-skin contact found that babies who received early skin-to-skin contact with their mothers during the first few hours after birth were more likely to breastfeed, and breastfed for an average of forty-three days longer than babies who didn't receive such contact. They also cried less, were warmer in the twenty-four hours after birth, their heart and breathing rate was around three beats and breaths per minute slower, and their blood-sugar levels were more stable than babies that didn't receive skin-to-skin contact.

The fact that babies who receive skin-to-skin contact are warmer doesn't seem just to be down to the transfer of heat from their mother's skin. A recent study found that some babies who were held to the skin within two hours of birth still had warmer feet, twenty-three hours later, than babies who spent their first hours in a cot. One explanation is that being held close to Mum reverses some of the effects of stress caused by the birth. Stress tends to divert blood flow away from extremities such as the feet, which would make them colder, and causes more of the baby's energy reserves to be used up, which might help to explain why babies who don't receive early skin-to-skin contact have lower blood-sugar levels.

Although it's not particularly surprising that newborns who are separated from their mothers cry more, what's interesting is that the type of cry they make is remarkably similar to a specific distress call made by other newborn mammals. One study that looked at crying in newborns found that it was virtually absent in babies who were in close contact with their mums, while those in cots tended to cry in short pulses of seven to forty-two seconds, separated by silent periods. When the researchers analysed tape recordings of these cries, they found that the pattern was very similar in different babies and it stopped as soon as they were united with their mothers. It was also distinct from other types of cry, such as those signalling hunger or pain.

One reason why babies might be separated from their mothers after birth is to give women some time to rest and recuperate, but if it's peace and quiet you're after, then it seems the best solution may be to hold the baby close to you – at least in these very early days.

Skin-to-skin contact may have other benefits for Mum, too. The pushing movements the baby makes with its legs stimulate the uterus to contract, helping to seal off the flow of blood as the placenta begins to peel away from the uterus and is finally delivered. Breast tissue also becomes profoundly sensitive to touch just before birth and any contact with it seems to trigger a surge of oxytocin, a hormone well known to promote social bonding. Previous studies in other mammals have suggested that newborns deliberately knead or paw the udders, and this stimulates the production of both oxytocin and milk.

Video recordings of newborn babies placed on their mother's naked chest suggest that they may do something similar. Around six minutes after birth, a newborn baby opens its eyes and starts to look around. Five minutes later, its hands, which were initially open and relaxed, begin to perform rhythmic, massage-like 'milking' movements on its mother's breast, which increase in frequency over the coming hour. Twenty minutes after birth, the baby begins to root or make sucking motions with its mouth, and shortly afterwards it begins to touch its mum's nipple, which becomes more erect as a result. Now the baby begins to lick the nipple, all the while continuing to massage its mother's breast. Finally, around eighty minutes after the birth, the newborn baby puts the nipple in its mouth and begins to suck.

The researchers who recorded these sequences took blood

samples from the mothers at regular intervals and found that their levels of oxytocin rose as the frequency of the baby's hand movements increased. Besides promoting bonding between Mum and baby, this may also help explain why early skin-to-skin contact is linked to higher rates of breastfeeding – many researchers think that the first hours after birth represent a critical period in which the connections between breast stimulation and the brain become reinforced. Others have noticed that levels of feel-good chemicals called endorphins in the blood double during breastfeeding, which might also promote bonding.

So what should parents do if Mum is unable to have early skin-to-skin contact with her baby – perhaps because of a medical emergency? Mums who have to be separated from their baby for prolonged periods can try massaging their breasts to try to mimic the baby's actions. Getting Dad to strip off and cuddle the baby may also have some effect (and I bet most dads would relish the chance to bond with their newborn, too). In 2007, Swedish researchers set out to investigate how babies respond to immediate skin-to-skin contact with their dads when their mothers were unable to hold them. It was a small study of twenty-nine babies, around half of whom had immediate skin-to-skin contact with their dads, while the rest were placed in cots beside them. Within fifteen minutes, the babies held by their dads had dramatically reduced their crying, and after sixty minutes they had entered a calm and drowsy state, while the babies lying in cots took around 110 minutes to reach the same state.

Sticking the baby on Dad's chest may also kick-start his paternal instinct: men release oxytocin in response to cuddling and interacting with their babies, and the more oxytocin they release, the more time they seem to want to spend with them (see 32: 'Do men change when they become dads?').

84

How do newborns know to seek the nipple for food?

BABIES ARE BORN with an acute sense of smell and early studies hint that they may use this to guide them towards their one and only food source: milk. In 2001, researchers conducted a fascinating experiment. Twenty-one newborns ranging from thirty-six hours to four days old were placed on to a warm bed, and either a clean cotton pad or one that their mothers had been wearing next to their breast for several hours was put in front of them, just out of reach. Using their legs to propel them forward, the babies began to inch their way towards the pads. Within three minutes, eighteen of the babies had inched towards the pad that smelled of the breast, compared to just three of the babies when they were given the control pad to examine.

Many women may notice that the area surrounding their nipples becomes increasingly bumpy during pregnancy and that these bumps, called Montgomery's tubercles, sometimes secrete an oily fluid. It was previously assumed that this fluid helped to keep the nipple lubricated, but Benoist Schaal and his colleagues at the National Centre for Scientific Research in Dijon, France, recently found that it may also help guide the baby to the breast. When they collected these secretions and exposed three-day-old babies to them, they found that the babies made more sucking and licking movements than when they smelled other scents, such as sweat or breast milk. Schaal also found that women with more than nine glands per breast started lactating around ten hours sooner than women with fewer glands, while their babies also fed more frequently and gained weight more quickly.

It's not only babies that may be sensitive to these odours. Another recent study found that smells associated with breast-feeding may put other women in the mood for sex. Breastfeeding women were asked to wear cotton pads next to their breasts and in their armpits for eight hours. A separate group of women were then given pads worn by the lactating women or unscented pads and asked to wipe them over their upper lip each morning and evening for three months, as well as keeping a diary of their sex lives and levels of sexual desire.

The women given pads worn by breastfeeding women experienced an increase in sexual desire and sexual fantasies, particularly during the second half of their menstrual cycles, when desire usually falls. Whether the chemicals responsible for this aphrodisiac effect are the same as those that newborns use to find the breast is unclear – neither has anyone tested whether men get similarly aroused by the smell of female breasts – but you might want to think more carefully about how you dispose of those used breast pads in future.

85

What causes the Moro or 'startle' reflex?

YOUR BRAND-NEW BABY is lying peacefully in his or her cot, when you accidentally bump against its sides. Melodramatically, the baby arches its back and flings its legs and arms out to the side with fingers spread, as if someone's just poured a bucket of cold water over its head.

Babies are born with a number of reflexes – actions performed in response to a stimulus, without conscious thought – but the

Moro reflex is probably the most dramatic (and amusing) of them. It is triggered if you bang something against the surface the baby is lying on, or if the baby's head drops slightly below the level of its body. Some have proposed that in our distant past the Moro reflex helped babies to survive by prompting them to cling on to and embrace their mothers if they felt as if they were falling.

Another reflex that could have played a similar role is the grasp reflex, which you can test by placing your finger in the centre of a newborn's palm: it should grip it tightly enough to hold even its own body weight. Back in 1891, Louis Robinson tested how long newborn infants were able to cling on to a horizontal rod with both hands using their grasp reflex. After testing sixty babies under the age of one month, he concluded that most could hang on for at least ten seconds, while one plucky infant managed to hang from the rod for two minutes and thirty-five seconds before letting go.

Monkeys also have these reflexes, which supports the idea that they may have been important in our evolutionary past. The grasp reflex is even stronger in baby monkeys, which often cling to their mother's coats as they are carried around, and tests have shown that they can hang one-handed from a horizontal rod for between seven and thirty-three minutes.

However, although apes and monkeys also possess the Moro reflex, this doesn't necessarily protect them against falls. One problem is that the main feature of the Moro reflex is the throwing out of the arms, rather than clinging on and embracing something – although neither babies nor monkeys produce the Moro reflex if they are grasping on to something at the time.

A group of Japanese researchers recently proposed that, rather than using the Moro reflex to break its fall, a baby may use it to catch its mother's attention if she starts to drop it. If a baby flings its arms out dramatically, Mum would immediately grab on to it and prevent its fall.

There are several other interesting reflexes you might want to look out for as well:

Rooting reflex

If you stroke your baby's cheek, they should start to turn towards your finger; this may help babies to find the nipple. This reflex disappears after about three months.

Tonic neck reflex

If you turn your baby's head to one side while they're awake and lying on their back, they should extend the same arm in front of their eyes and flex their other arm; this might help prepare the baby for reaching. This reflex disappears after about four months.

Stepping reflex

If you hold your baby under their arms with their feet touching the floor, they should start lifting their legs in a stepping motion; this may help prepare them for walking. This disappears after around two months.

Babinski reflex

If you stroke the sole of your baby's foot from toe to heel, you will probably see their toes spread out and curl as the foot curls in; the purpose of this reflex is unknown. It disappears at eight to twelve months.

Swimming reflex

Dunk a baby face down into a pool of water and they'll instinctively hold their breath, paddle and kick in a swimming motion. Take care with this one, though, as they may still swallow large amounts of water. It may have evolved to help babies survive if they fall into water, but it disappears at four to six months.

Baby Bodies

86

Are baby growth charts accurate?

It's a moment many mums come to dread: going to the baby clinic to get their baby weighed and being told they aren't growing fast enough, or indeed that they're surprisingly big for their age (even though they look and seem perfectly healthy).

In the UK, growth charts for height and weight are included in every baby's personal health record, a book that serves as the main record of a child's health, growth and development during the early years of its life. Every few weeks your baby will be weighed and its measurements plotted against a standard curve, which tells you roughly how it compares to other babies of the same age. Nine lines snake their way across a standard growth chart, each representing a 'centile' above or below the average. If your baby's weight lies on the ninth centile, this means that for every one hundred healthy babies of the same age, nine would be expected to weigh less and ninety-one would weigh more than that child.

Growth charts can be a great source of anxiety for new

parents, particularly as babies can jump up and down centiles within a matter of weeks. My daughter was on the twenty-fifth centile for the first four weeks and then slipped down to the ninth centile, before starting to climb up towards the fiftieth. Each time I was told not to worry – with the caveat that if she continued on the same trajectory then it might be time to start getting worried.

More experienced mums informed me that growth charts are inaccurate because they're based on measurements of bottle-fed babies, who tend to be heavier, although the truth is that this depends on the country you live in and the charts being used. But this argument is also fast becoming redundant, because the UK and many other countries, including the US, recently replaced their old charts with growth charts provided by the World Health Organization (WHO), which are based on data from exclusively breast-fed babies.

You'll know if the chart your doctor or health visitor is using is one of the new WHO charts because they have a separate section for preterm babies, no lines between 0 and 2 weeks, and the fiftieth centile (the middle line) is no longer emphasized in bold as it was on the older charts, so parents feel less that this is the ideal weight which all babies should be striving to attain.

The WHO introduced the new charts in 2006 after collecting data from babies born in six different countries. Because exclusive breastfeeding is now considered the gold standard by the WHO and many other health bodies, the idea behind the charts was to describe growth as it should occur under optimum conditions rather than as it actually occurs in the general population.

However, the introduction of the new charts has not been entirely without controversy. Compared with existing growth charts from the US and the Netherlands, and a multinational growth chart representing other European countries, babies measured against the WHO charts are judged to be heavier during the first six months of life and lighter between six and twenty-four months of age. In other words, a four-month-old baby previously classified as healthy might now be classified as underweight, while a twelve-month-old previously thought to be a normal weight might now be judged as overweight.

This has come as a surprise to many, as these previous charts included some bottle-fed babies, which are generally assumed to be heavier. One explanation is that the criteria used to decide which babies should be included in the WHO study were so strict that they have led to a skewed picture of what weight gain should look like in breastfed babies. If the WHO standards really are too high, this could serve to discourage women from exclusive breast-feeding, or to give up altogether. 'If the child's weight is towards the low end of the range a mother may conclude that her milk supply is inadequate,' says Ekhard Ziegler, Professor of Pediatrics at the University of Iowa in Iowa City.

However, Tim Cole, a statistician at the MRC Centre of Epidemiology for Child Health in London, who helped adapt the WHO charts for the UK, says the patterns aren't necessarily suspicious. Although the common perception is that bottle-fed babies are heavier, 'Breastfed babies do grow rapidly to start with and then slow down,' he says. In the UK, babies are actually more likely to be classified as overweight using the new charts compared to the older ones, regardless of their age.

So what message should new parents take from all this? If even the experts can't agree on the best way of assessing a baby's weight, perhaps we should all relax a bit and stop worrying about how our baby's growth is tracking compared to the average. So long as they're not completely off the charts, the chances are they're fine.

Growth charts are also supposed to reflect the fact that

individual babies differ in height and build, so there's nothing wrong with your baby being in the first centile or the ninety-ninth centile. Another common misunderstanding is that regardless of how heavy a baby is, it should broadly stick to the same growth curve and not switch from one centile to another.

A baby's birth weight doesn't just reflect the weight it is genetically programmed to be but how good the mother's uterus and placenta were at supplying the baby with nutrients and how much the mother ate during pregnancy. But, once the baby is born, these other factors cease to be relevant. This means that babies may lose or gain a fair amount of weight while they are finding their natural balance. 'Particularly during the first year, babies cross centiles a lot,' says Cole. 'Generally, it has to be pretty extreme before people should start worrying about it, particularly during the first six months.' If you are worried about the rate at which your baby is gaining weight, you can buy a set of acetate 'thrive lines' from the UK's Child Growth Foundation that can be placed over growth charts and will tell you if there is genuine cause for concern.

Neither should parents feel pressured into having their baby weighed every time they see a health visitor or midwife. In fact, the latest UK recommendations say that all babies should be weighed at six to eight weeks, and then only when there is parental or professional concern. Even where there is concern, babies shouldn't be weighed more than once a month before the age of six months, or every two months between six and twelve months. Weighing babies more frequently than this can lead to misleading results, because natural fluctuations in weight and differences between weighing scales may be greater than any overall weight gain during that period. Weight gain also slows down as babies approach their first birthdays.

Finally, when it comes to really young babies, the growth charts should be chucked out altogether. Generally, the most important consideration during the first two weeks of a baby's life is not its position on a chart relative to other babies, but how its current weight compares to its birth weight. Most babies are expected to lose a bit of weight after birth, but they usually regain their birth weight by two weeks, and only 3 to 7 per cent of babies lose as much as 10 per cent of their birth weight. Even then, at least one study found no major medical problems among babies that lost more than 10 per cent of their birth weight in the first twelve days. Unless doctors have a genuine medical concern about a baby's weight loss, parents shouldn't feel pressurized into feeding their baby supplemental milk, particularly if this could interfere with the establishment of breastfeeding.

87

Do big babies grow into big adults?

IT IS CLEAR that tall parents tend to produce tall children, but factors such as putting on a lot of weight or smoking during pregnancy and gestational diabetes are known to impact on a baby's size at birth. Mothers seem to have a greater influence over a baby's growth during pregnancy than fathers, probably as a means of safeguarding against serious birth complications (see 16: 'Do big parents have bigger babies?').

That said, several studies have found an association between a baby's length and weight at birth and the size it ultimately grows into as an adult. The strongest correlation seems to be for length. As a rough guide, one recent study from Denmark found that baby boys who were shorter than 47cm at birth grew into men with an average height of 175.2cm, while those longer than 56cm had an adult height of 184.3cm. Unfortunately, it didn't look at baby girls' heights.

A separate study found that birth weight and length make independent contributions to the eventual size of a person. So if your newborn is both long and heavy, you may well have a future giant on your hands.

88

Why don't babies have moles?

THE MAJORITY OF babies are born with flawless skin, but moles and freckles will appear as they progress through childhood and adolescence – usually from the age of around six months. By the time they are adults, most will have around twelve to twenty perfectly harmless moles on their skin.

Exactly what triggers mole growth is unknown, but it is likely to be an interaction between genetic factors and sun exposure. Moles tend to appear on skin that is regularly exposed to the sun, and they usually start off as a flat, round spot that will grow evenly on both sides as the child gets older. Sometimes the mole may start to grow into a raised bump, and sometimes it may get lighter or darker, but unless the mole looks itchy, bleeds or is different to other moles, it is unlikely to be anything to worry about.

Many studies have shown that moles are more common in children with fair skin or those who are frequently exposed to the sun, so if your child does play outside a lot, then keeping their skin covered up with clothing should reduce the number of moles that appear. How many moles a child will develop is also inherited to a certain extent, so if you and your partner have a lot of moles, your baby is likely to develop more too. While a few moles are nothing to worry about, having lots of moles (more than a hundred when you're an adult) seems to put you at increased risk of skin cancer. Going on holiday to hot countries also seems to increase a child's chances of developing moles.

Around 1 per cent of babies are born with a mole, which in this case is a type of birthmark. Most are harmless, but if your baby is born with a really big birthmark (more than 20cm in diameter), they may be at a slightly increased risk of developing a form of skin cancer called melanoma when they get older.

89

What causes colic?

COLIC IS MOST parents' vision of hell. All babies cry, some more than others, but colic is defined as an otherwise healthy baby showing periods of intense or unexplained crying lasting for more than three hours a day, more than three days a week, for more than three weeks. It tends to peak at around the age of six weeks, is usually worse in the late afternoon and evening, and often suddenly stops at the age of two to three months. Around a fifth of two-month-olds suffer from colic, and it can have serious consequences. Mothers of colicky babies are more likely to suffer from postnatal depression, more likely to give up breastfeeding, and colic occasionally prompts some to do the unthinkable: to lash out and harm their child.

For thousands of years, doctors have blamed excessive crying on problems related to digestion – indeed, the word 'colic' comes from the Greek *kolikos*, which translates as 'of the colon'. Yet it is often used as a catch-all phrase to diagnose something that may have multiple causes. Current theories include food allergy, an immature digestive system, that it's part of normal brain development (albeit at the extreme end of the spectrum), and the idea that some babies are simply born with more difficult temperaments.

Some recent research has suggested that colic may be down to low-level inflammation of the baby's gut or painful muscle spasms (a bit like inflammatory bowel syndrome in adults). There is also some evidence that herbal teas containing vervain, camomile, fennel, liquorice or lemon balm can help colic by relaxing the muscles in the gut, but giving babies too much extra liquid can reduce their appetite for milk. A drug called Simethicone (marketed in the UK as Infacol®) claims to reduce the surface tension of trapped air bubbles in the gut, enabling gas to be expelled more easily. Although one trial found that Simethicone reduced the number of crying attacks on days four to seven of treatment compared with placebo, it included only twenty-six babies and provided no details on how colic was defined. Two other studies found no significant effect.

Some have suggested that colicky babies simply have 'difficult' personalities, but most experts agree that this is unlikely to be a major cause of colic (although it may be a contributing factor). And, while there may be a link between colic and anxious mothering, it's probable that this is not a direct cause. However, parents should try to watch for early signs that their baby may be tired or hungry, as these things can make crying worse.

Many paediatricians believe that so long as babies are growing and developing normally, there is no need to treat colic, because it usually disappears with time. The trouble with this is that some babies labelled as having colic may have an underlying medical problem that could be treated, potentially ending their suffering.

In one small study of twenty-four healthy babies and nineteen babies diagnosed with colic, researchers found that up to half of those in the colic group showed evidence of feeding problems, including less rhythmic sucking and problems coordinating sucking, breathing and swallowing. In addition, ultrasound scans revealed more evidence of gastro-intestinal reflux in the colic group – something that was backed up by reports from mothers of colicky babies.

Before simply swallowing a diagnosis of colic and leaving it to run its course, you should try to eliminate some of the following causes:

Gastro-oesophageal reflux

Spitting up milk is extremely common among newborn babies and is often nothing to worry about. Reflux doesn't usually cause pain, as spit-up is actually pH neutral for two hours after a feed because milk acts as a buffer to stomach acid. However, if your colicky baby is also spitting up blood, refusing to feed, failing to gain weight, or projectile vomiting, these could be signs of more serious gastro-oesophageal reflux disease and you should consult a doctor.

Infection

Infections are sometimes mistakenly labelled as colic, because they can also trigger inconsolable crying. One study of 237 babies taken to hospital because they wouldn't stop crying found that

5 per cent of them had an underlying illness, most commonly a urinary-tract infection.

Breastfeeding problems

Failing to properly latch on to the breast can lead to excessive crying because the baby isn't getting enough milk. Classic warning signs during the first six weeks include producing fewer than six wet cloth nappies or four heavily wet disposable nappies per day, and fewer than three to four yellow stools per day (although babies do vary in how often they produce stools).

Functional lactose overload

Babies that don't feed for long enough to get through the relatively watery fore-milk to the creamier hind-milk can develop something called functional lactose overload. Fat levels in milk increase as the breastfeed progresses, and this fat slows down the progress of milk through the intestines, giving it more time to be digested. But if a baby drinks only the watery fore-milk, this passes through the gut too quickly and the undigested protein then ferments in the colon. The result is frothy and explosive poos, a bloated belly, crying and hunger. One study of seventy-seven five-week-olds found that those with colic had lower levels of a hormone that's released in response to eating a high-fat meal, suggesting that the milk they were consuming wasn't rich enough, probably because they weren't taking long enough over each feed. A different study which compared the effect of either offering one breast at each feed or both breasts for a shorter period of time found a lower incidence of colic in the babies that fed from just one breast, as well as less engorgement and mastitis among the mothers.

Food allergy

Occasionally, a baby may be allergic to something in its mother's diet, but despite suggestions that mums should stop eating soy, wheat, nuts or fish to ease their baby's discomfort, the only thing that has consistently been linked to inconsolable crying is an allergy to cows' milk. Between 5 and 15 per cent of babies show symptoms suggesting they may be allergic to some of the proteins in cows' milk, but only half of these are actually thought to suffer from it. Key symptoms to look out for include crying, a runny nose and persistent cough, a skin rash or eczema, frequent regurgitation of milk, and diarrhoea or constipation. Cutting dairy products out of your own diet for two weeks should tell you if an allergy to cows' milk is at the root of your baby's problems. If the baby is formula-feeding, you can try switching to an extensively hydrolysed or hypoallergenic formula. According to European experts, there is no evidence supporting the use of formula containing soy protein in babies suffering from colic, regurgitation or prolonged crying.

Lactose intolerance

An allergy to cows' milk is different to lactose intolerance, where babies don't have enough of the lactase enzyme that usually breaks down a sugar found in milk called lactose. The result is too much of this sugar entering the colon, where it ferments, causing bloating and pain. Several small studies have found that babies suffering from colic have more hydrogen gas in their breath, which is a product of lactose fermentation. Likewise, a few early studies have suggested that feeding babies formula treated with lactase enzyme can ease symptoms of colic in some. However, genuine lactose intolerance is pretty rare during the first year of life and some paediatricians feel that it is being over-diagnosed. More likely is functional lactose overload or an allergy to cows' milk, although babies may develop temporary lactose intolerance as a result of damage to the gut lining caused by gastroenteritis or an allergy to cows' milk. Feeding a baby with an allergy to cows' milk a lactase-free formula may make things worse if the formula contains cows' milk protein, as it will continue to irritate the gut.

90

What is the best way to settle a crying baby?

ANYONE WHO HAS been caught in an enclosed space with a howling baby can testify to how stressful it can be. Frantic parents will try anything: rocking their baby, doing endless laps of the park with the buggy, even shelling out hundreds of pounds on cranial osteopathy, vibrating mattresses or devices that claim to interpret a baby's cries and tell you if it is hungry, bored, sleepy or uncomfortable.

Although there will undoubtedly be babies that respond to each of these different strategies, scientists have also weighed in on the issue and put some of them to the test.

One of the first strategies parents often turn to is carrying their baby around with them, often in a sling or wrap. Studies have reached mixed conclusions about whether or not this works: some found that it is effective so long as you start carrying before the age of six weeks, while others found no reduction in total crying as a result of carrying. There are also concerns that prolonged carrying of a baby simply teaches it how to fall asleep in its parents' arms, rather than on its own in its cot.

Certainly carrying doesn't seem to be particularly helpful for babies with colic. Even in societies where babies are constantly carried around by their mothers, there are babies who continue to cry inconsolably. And a randomized controlled trial of sixty-six mothers who said their babies cried excessively found no difference in the duration or frequency of fussing and crying between those who simply rocked and soothed their babies and those who increased the amount of time they spent carrying their babies by around two hours per day.

One thing that does seem to reduce crying is introducing a flexible routine to a baby's waking hours, similar to the kind of approach suggested by 'Baby Whisperer' Tracy Hogg. One study found that a repeated pattern in which naps were followed by a feed, then some interactive playtime, then some quiet time alone on a mat or in a playpen during which parents watched for any signs of tiredness (such as eye-rubbing) and put the baby back down for a sleep (either swaddled or tucked in tightly with sheets and blankets) reduced crying in babies under eight weeks of age.

Other studies have suggested that swaddling (or wrapping) the upper half of the body to restrict a baby's arm movements also seems to increase the amount of daytime sleep they enjoy, although swaddled babies should never be placed face down, and the swaddling shouldn't be so tight that babies can't move their limbs at all. Skin-to-skin contact can reduce crying in very young infants, and there is some evidence that massage can reduce crying and boost sleep and relaxation in babies under six months of age.

If you've tried everything, can find nothing wrong and your baby still won't stop crying, most paediatricians agree that you should put your child down somewhere safe and take a break. Sadly, a minority of babies seem prone to crying for no apparent reason. The good news is that most babies that cry a lot at the age of five to six weeks are no more likely than quieter babies to continue being unsettled by the age of three months, and many will also start sleeping through the night at this age.

91

Why do some belly buttons become 'outies'?

BELLY BUTTONS ARE the shrivelled-up remnant of the umbilical cord, but just why some of them end up sticking out when others form a deep hole remains something of a mystery. One theory is that the 'outieness' is influenced by the amount of scar tissue that forms underneath and around the severed umbilical cord. 'Outies' can also form because the opening to the cord doesn't close up properly, resulting in something called an umbilical hernia. Such hernias are usually painless, are more visible when a baby is crying or has a particularly full belly and will often disappear by about eighteen months of age.

There's currently no way of predicting whether your baby will have an outie. Around 20 per cent of newborns have them, but many will turn into innies as the babies age and fat accumulates around the stalk of the scar, pulling it inwards. Malnutrition can change the shape of the hood on top of the navel, while the size of the belly button is thought to relate to the size of the umbilical cord during pregnancy and may therefore signal how many nutrients the foetus received

Aki Sinkkonen of the University of Helsinki in] researched the evolution of belly buttons and believ provide a signal about a woman's reproductive fitnes no one has done an equivalent study of male belly buttons). Studies suggest that the most desirable shape for a female belly button is a symmetrical T or vertical shape, with a large hood of skin at the top, and neither excessively big nor too small. However, Sinkkonen says that 'outies' are common and socially appreciated in sub-Saharan Africa.

92

Do babies heal faster than adults?

INJURIES ARE A fact of life, but the first time you spot blood oozing from your baby's perfect peachy skin is a traumatic event for any parent – especially if the injury is your own fault. The first time it happened to me, I was trimming my month-old daughter's fingernails and accidentally nipped the skin, which produced an obscene amount of blood for such a trivial injury.

The miraculous thing is that babies often seem to heal with little or no scarring. Injure an adult's skin and the body releases chemicals which promote swelling and inflammation, rallying immune cells to the scene. As the skin begins to heal, cells secrete collagen fibres to patch up the wound, but whereas in normal skin these collagen fibres form a criss-cross pattern, in scar tissue they are laid down in parallel bundles, making the tissue less flexible. In contrast, the immune system of the foetus is still immature and there is very little inflammation when the skin is injured, which seems to allow it to heal better. Even when relatively major operations are performed on foetuses in the uterus, they can be born blemish-free.

This superior healing ability also seems to extend beyond

birth. In a recent study, doctors at the Children's Hospital of Philadelphia analysed samples of human foreskin removed during circumcision. They were curious to know why it is that the amount of scarring and the number of complications they see following routine operations to repair hypospadias – a common birth defect in which the hole at the end of the penis is in the wrong place – increase as children age.

When they analysed skin from babies younger than twenty-eight days old, they found that it produced far lower levels of chemicals known to promote inflammation than skin from six- to twelve-month-olds. The amounts of these chemicals were even higher in older children.

It's not just scarring. Recent studies have shown that if the tip of a newborn mouse's heart is surgically removed, within a matter of weeks it will have grown back. Further investigation revealed that the remaining heart cells started to divide and create new muscle tissue – something that would be unthinkable in adults. Although no one has yet proved that human babies have the same abilities, paediatric surgeons report that certain types of heart surgery are far more successful when performed within the first few months of birth than if they wait until the baby is older. If we can understand the complicated chemical signals that enable this regeneration to take place, we might one day be able to help adults achieve a similar feat.

93

Why do babies gnaw on things even before they start teething?

RED CHEEKS? DROOLING and irritable? Chewing on anything they can get their hands on? The baby has got to be teething, right? Many other symptoms are also attributed to teething, including diarrhoea, sleep disturbances and fever. But is teething really the villain it's made out to be?

Babies are actually born with their milk teeth already sitting in a neat row beneath the gum, but most babies don't cut their first teeth until about six months of age (although, in rare cases, babies are born with visible teeth, much to the dismay of mums who had intended to breastfeed). First to appear are the central (cutting) incisors, followed soon afterwards by the lateral incisors that sit next to them, then the first molars and the canines. Finally, at around the age of two, the second set of molars starts to appear. Girls tend to get their teeth earlier than boys, although the reason for this is unknown.

Just what causes teeth to erupt from the gum remains a medical mystery, but we do know that teeth aren't pushed upwards by the growth of the root or the jaw bone. The tooth itself seems to be a passive bystander in the process, but the structure from which it emerges in the jaw is capable of producing all sorts of chemicals which can prompt cell growth and inflammation, and these could theoretically trigger the irritation, drooling and vivid red cheeks so often considered a sign of teething. Indeed, some of these chemicals have previously been linked to fever and to disturbances in sleep and appetite.

Although several studies have suggested that babies who are

teething show more mouthing, sucking, drooling and fever than non-teething babies, these studies suffer from the fact that the people making the observations knew they were participating in a study about teething and may therefore have been biased.

To investigate further, Melissa Wake and her colleagues at the Royal Children's Hospital in Parkville, Australia, asked the parents of twenty-one babies attending day-care centres to fill out daily questionnaires describing their child's mood, wellness, drooling, sleep, the contents of their nappies and any rashes or facial flushing they'd noticed over the previous twenty-four hours. The parents and nursery staff dutifully completed these questionnaires for seven months, and every day a dental therapist came to take the babies' temperature and check them for any signs of tooth eruption.

Contrary to widespread belief, the researchers found no evidence for increased illness or fever, mood or sleep disturbances, drooling, strong urine, red cheeks or rashes in the five days leading up to the eruption of a tooth and on the day itself – although parents did report looser stools in the five days leading up to the appearance of a tooth.

More recently, Brazilian researchers conducted a similar study of forty-seven babies being cared for by their mothers at home. They found some evidence of increased sleep disturbance, drooling, rash, runny nose, diarrhoea, loss of appetite and irritability on the day of tooth eruption itself, and one day afterwards. However, these symptoms weren't present in the days running up to it, meaning there was no way of predicting when teeth might put in an appearance.

So why are these findings so different from what we have been led to believe? Wake says that teething may have become a scapegoat for many other events occurring between the ages of six and twenty-four months, including a sudden rise in respiratory, middle-ear and diarrheal infections, as well as other developmental processes that could cause babies to be more irritable than normal or to sleep less.

If a baby is being fussy and irritable, it's often easier to attribute it to a known physical cause than to accept that they're simply in a bad mood – or that there is a behavioural problem that should be addressed. I know my own daughter has generated far more sympathy from me if she has been whiny and difficult and sporting red cheeks, compared to when there is no obvious physical cause.

94

Why is baby hair a different colour to adult hair, and why do babies lose their hair?

SOME BABIES ARE born bald, while others are born sporting a full head of hair that often bears no relation to the colour it will be a few months later (let alone when the child becomes an adult). My daughter, Matilda, was born semi-bald but with a fringe of brown hair around the base of her scalp that made her look a bit like a monk. A month later this fell out and was replaced with a mop of silky blonde hair. Another friend's daughter was born with bright-orange hair, but she too is now blonde.

A baby's hair begins to grow from around twenty weeks of pregnancy, and the first hair they produce – vellus hair – tends to be silky and fine, and is similar to the hair that grows on the rest of our bodies. As the hair follicles on the head grow larger, this early hair begins to be replaced with intermediate hair, although some of this early baby hair continues to grow. Finally, at around the age of two, intermediate hair is replaced by thicker terminal hair that is more similar in texture to adult hair.

A pigment called melanin plays a large role in determining hair colour, and there are two types: eumelanin, which colours hair brown to black, and pheomelanin, which colours hair blond to red. In general, your hair colour depends on the proportion of these pigments that your body possesses, but many people will have noticed that blond children often end up with darker hair as adults. This is because the enzyme that makes these pigments becomes more active as people age, and under the influence of hormones that start being produced at puberty.

Although no one has tested this, it could well be that the hormones coursing through Mum's body during pregnancy mean that a baby's first hair is a different colour to their later hair. We know that high levels of oestrogen from the mother's body sometimes cause baby girls to have swollen breasts and even produce milk or menstruate during the first days after birth.

So what causes some babies to lose their hair? Hair usually starts growing at the front of the head and spreads backwards as the baby gets older. It also runs in cycles, consisting of a growing phase, an intermediate phase and a resting phase – after which the hair falls out. How much hair a baby is born with often depends on what stage of this cycle their hair follicles have reached. Some babies will have lost all their vellus hair by the time they are born (and even started growing new hair), while others will still be left with a bit of vellus hair – usually towards the back of the head, like Matilda was born with.

These growth cycles can also overlap to some degree, which means that babies can have a mixture of different hair types at the same time. As a general rule, babies with darker skin tend to have longer cycles of hair growth, so that they are often born with fuller heads of hair than fair babies.

One patch of hair towards the back of the scalp has a stubborn tendency not to enter its final resting phase until around three months after birth. Many parents notice a bald patch of skin on the back of their baby's head at around this time, which is usually blamed on babies rubbing their heads against their cot mattresses. Although rubbing the area probably helps the hair to fall out, this bald patch is actually caused by the shedding of this final patch of vellus hair. The phenomenon also has a catchy name: 'occipital neonatal alopecia'.

95

Can newborns sweat?

BABIES ARE CAPABLE of regulating their internal body temperature from birth (unless they're very premature), although they're not as good at it as adults. One of the mechanisms they use is sweating, which causes heat to be lost by evaporation. Studies have shown that most babies born after thirty-six weeks are capable of sweating from day one, although they don't sweat as much as adults and may need higher temperatures to trigger it. This means that sweating isn't necessarily a good guide to how hot a baby is, and you're better off feeling the back of their neck or chest to get an indication of their temperature.

96

Do fontanelles hurt?

THE SKULL IS made up of seven bones, but these aren't joined up at birth. This means a baby's head can change shape and be slightly compressed as it makes its way through the birth canal, making it easier for the baby to be born. The areas where these bones join are called the fontanelles (or soft spots), although there are only really two which are big enough for you to feel. The most obvious one is located on the top of the head, feels like a soft diamond and will remain there for up to nineteen months, although you may feel another small one at the back of the head, which usually closes after two months. As well as enabling the head to be compressed, the fontanelles also allow the brain some room to grow, before the bones of the skull eventually join up and harden.

It's difficult to know how hard you'd have to press the fontanelles for it to hurt your baby (and I doubt many would want to try). Although clearly not as solid as bone, they're relatively robust, as the brain is protected by a thick fibrous membrane that stands between it and the skin.

97

Are dummies good or bad for babies?

THE FASHION FOR dummies waxes and wanes. Right now in middle-class Britain, there seems to be a general tut of disapproval at any baby spotted with a dummy in its mouth, even though many a grandmother swears by them to soothe crying babies and settle them to sleep. Even so, an estimated 75 to 85 per cent of babies in Western countries use a dummy – at least in private.

So who is right? There's no doubt that young babies love to suck (be it on a thumb or on a piece of moulded plastic) and paediatricians widely regard sucking as one of the best ways of relieving pain in young babies.

The evidence for dummies helping to ward off sudden infant death syndrome (SIDS) is also pretty compelling – to the extent that the American Academy of Pediatrics now recommends offering a dummy at nap-time and bedtime to reduce the risk of SIDS. Studies have estimated that one SIDS death could be prevented for every 2,733 babies who use a dummy when put down to sleep. There are several theories as to why this might be. One is that babies sleeping with a dummy seem to be easier to arouse, so might be more likely to wake up if something happens to them during the night. Another theory is that a dummy makes it easier for a baby to breathe through its mouth if its nose becomes blocked. However, paediatricians do not recommend forcing a baby to use a dummy if it doesn't want to, or reinserting it if the baby spits it out.

The risk of SIDS is greatest in babies up to the age of six months, so after this the health benefits of using a dummy start to fade. One reason to try to wean babies off them after this point is a slightly increased risk of other infections. Several studies have found that dummies are often colonized with yeast and bacteria and latex dummies seem to allow more of these organisms to flourish than silicone ones. Sucking a thumb or fingers can also increase a baby's exposure to such organisms, although no one has directly compared whether this is any better or worse than dummy-sucking.

While most of these organisms won't cause disease, a study of more than 10,000 babies in the UK found that babies who sucked either a dummy or their fingers at the age of fifteen months had a higher incidence of earache and colic compared with those that didn't – although it's possible that dummies were being used to calm babies that often got sick for different reasons. The effect is likely to be small, though. Another study of 476 children found that 35 per cent of those who used dummies developed at least one middle-ear infection, compared to 32 per cent of those who didn't suck, while rates of recurrent ear infection were 16 per cent in the dummy group, compared to 11 per cent among controls.

Finally, many parents worry about the effect that dummies or thumb-sucking might be having on their child's mouth development. Both the American Dental Association and the American Academy of Pediatric Dentistry recommend that dummies be discouraged after four years of age because of an increased risk of malocclusion, or misalignment of the teeth. In one study, the prevalence of malocclusion was around 71 per cent in children who used a dummy or sucked their thumb for more than four years, compared with 32 per cent in children who stopped sucking between the age of three and four, and just 14 per cent in those who stopped sucking before the age of two.

Sleep

98

Can babies tell the difference between night and day?

NEWBORNS TEND TO sleep a lot – at least for the first couple of weeks – which often comes as a relief to their shell-shocked parents. The average newborn crams in around sixteen to eighteen hours of sleep per day, which tends to drop to around fourteen to fifteen hours by the time they are a month old. At this stage, a newborn's sleep patterns are more influenced by the fullness of its belly than by whether it's light or dark outside, but by two or three months, it is beginning to grasp the notion of night and day.

As anyone who has suffered from jet-lag will know, we humans have a firmly entrenched body clock, which means our bodies respond differently depending on what time of day it is. As night falls, we produce more of the sleep-promoting hormone, melatonin, which makes us increasingly drowsy as the evening progresses.

So what about babies? The body's master clock is located at the front of the brain in an area called the suprechiasmic nucleus (SCN). It is in direct contact with the eyes and can be shifted forwards or backwards by exposure to bright light, which is how the body manages to reset itself within a couple of days when you travel across time zones. The SCN is thought to have developed by around twenty weeks of pregnancy, and foetuses do show differences in their heart rates, breathing movements and hormone levels depending on the time of day – although these rhythms are probably being driven by signals from the mother, rather than being self-generated.

The SCN continues gradually to mature after birth. For the first few weeks, a baby's sleep is distributed pretty evenly over twenty-four hours, but by six weeks most babies are slightly more active during the day and sleepier at night, and by three months, higher levels of melatonin can be detected in a baby's blood at night.

Your grandmother may tell you that a good way of getting a baby to sleep through the night is to stick them outside in the garden during the daytime. Although you may be suspicious that this is simply a ruse for the mother to get some peace and quiet, they might be correct: research has shown that two- to three-month-olds who are exposed to plenty of bright sunlight during the day sleep better at night. This is probably for the same reason that people overcome jet-lag faster if they get outdoors into the sunshine – bright light plays a crucial role in setting the body's central clock and ensuring that it runs on time.

99

Are a baby's sleep patterns inherited?

MOST PEOPLE BROADLY fall into one of two categories: 'larks', who wake up early and like to turn into bed soon after it gets dark, and 'owls', who like to slumber through the morning and are more active during the evening. Just which category you fall into seems to be inherited to some degree, which is good news for owl-like parents who were dreading being woken up by a 'dawn baby chorus'.

Interestingly, just how much of a lark or owl someone becomes also seems to be driven by the season in which they were born. Several European and Canadian studies have suggested that babies born during spring and summer are more likely to turn into owls, while those born in autumn and winter are more likely to become larks. The effect was particularly strong among babies born in Montreal during late September and October, a high proportion of whom became morning types. This might be explained by the fact that the months from October to December tend to be gloomy for those living in Montreal, with the lowest average hours of sunshine over the year. Researchers believe that there may be a critical window of time after birth when the body clock is extremely sensitive to light and its future pattern can be set. Studies in rats have suggested that those raised in dim light are more sensitive to its effects, finding it harder to sleep during daylight, for example.

What about other aspects of sleep? Much of what we know about the inheritance of sleep patterns comes from studies of identical twins. Because they share the same genes, you can look at differences in twins' behaviour to calculate what proportion comes down to genetics and what proportion can be attributed to external factors such as how they were raised, the hours they work and their social lives. Twin studies have suggested that approximately a third of the difference in the amount of sleep people need, the amount of time it takes people to fall asleep and how often they wake during the night can be explained by genetic factors. This means that although a baby's sleep patterns should show some relation to how you sleep as parents, external factors have a greater influence.

100

How can I get my baby to sleep through the night?

THE AVERAGE NEWBORN sleeps for sixteen to eighteen hours a day, but there is still a lot of variability between individuals. One large study found that some newborns slept for just nine hours per day, while others clocked up a whopping nineteen hours of sleep over twenty-four hours. The same is true at two months of age, with some babies sleeping for just twelve hours a day and others managing twenty-one.

The good news for parents who find themselves blessed with a particularly sleepy newborn is that the amount of sleep babies need and their general degree of fussiness seems to remain relatively consistent as they get older – although how often they wake, feed and cry can vary according to what's going on in their lives.

As for sleeping through the night, parents shouldn't expect too much too soon. Although studies have found that up to three-quarters of babies are sleeping through the night by twelve weeks of age, what 'sleeping through the night' means to researchers is five hours of undisturbed sleep in a row – a definition I suspect most parents would find reason to quibble with. A review of twenty-six studies focusing on babies' sleep found that just 37 per cent of healthy three-month-olds regularly slept for eight hours at night without waking their parents.

While the amount of sleep a baby needs is inherited to some degree, research suggests that parents' interactions with their baby play a crucial role in terms of getting them to sleep through the night. So how do you encourage your baby to fall into the sleepy category?

Most sleep researchers agree that introducing a consistent bedtime routine in the run-up to bed such as brushing teeth, taking a bath and reading a story helps babies to fall asleep, as does waking them up at the same time each morning (assuming they don't wake you up first) and ensuring that they get some kind of physical exercise during the day. Making a clear distinction between night- and daytime also seems to help, by settling babies for naps in rooms that aren't too dark during the daytime and minimizing interaction with the baby at night, for example. At least one study found that babies exposed to more natural light in the early afternoon slept better at night than those exposed to less.

Babies are creatures of habit and quickly learn to associate certain cues with sleep. This means that if parents aren't careful, they can also form bad habits. A review of studies on night-waking and other bedtime issues found that some of the most common problems included needing to be rocked, held or for the parent to be present in order for the baby to fall asleep, needing to be fed to sleep and habitually waking several times during the night. It is unlikely that parents set out to form such habits. More likely they are born out of desperation – taking a baby into bed and breastfeeding it back to sleep when it wakes for the fifth hour in a row, for example.

Trying to prevent bad habits from forming in the first place is often easier than trying to break them once they've become established. Most experts agree that parents should try to put babies down in their cots while they are still awake so the baby doesn't become reliant on them being there in order to fall asleep.

There are a few other tactics which parents can use to try to establish good sleep habits from a tender age. A study of 610 women found that those who were taught the following techniques when their babies were ten days old were 10 per cent more likely to have a baby that consistently slept for five hours or more at twelve weeks, compared to parents who received little or no advice.

(The next eight points are taken from 'Use of a behavioural programme in the first three months to prevent infant crying and sleeping problems', *Journal of Paediatrics and Child Health*, June 2001, vol. 37, issue 3, pp. 289–97.)

1 During the daytime, feed your baby whenever they seem hungry and spend plenty of time interacting with them. Try to bathe them at around the same time each day.

2 In the evening, encourage your baby to take a 'focal feed' – either a longer than normal breastfeed or a larger formula feed – between 10 p.m. and midnight.

3 If your baby is still awake after feeding and changing, try not to hold, rock or nurse them to sleep. Instead, put them into their cot while they're still awake and leave them to settle themselves. Lights should be dimmed, although it doesn't have to be completely dark.

4 If your baby doesn't settle, check the following:

(**a**) Does their nappy need changing?

(**b**) Do they need winding?

(**c**) Are they too hot or cold?

(**d**) Stroke and talk softly to your baby. If necessary, pick them up and have a cuddle.

Give each tactic at least ten minutes to work before moving on to the next one.

5 Try to distinguish between genuine crying and fretting in order to reduce the number of times your baby is picked up at night. Babies often fret before settling to sleep and don't always need further attention.

6 Try to keep the lights dimmed during night feeds. Respond to your baby's needs by changing their nappy or feeding them and then settle them back to bed. If your baby doesn't settle, work through step 4 again. Avoid playing or socializing with your baby.

7 If your baby is waking more than every three to four hours at night, they probably don't need their nappy changed every time.

8 Make nights as uninteresting as possible for your baby. Night-time is for sleeping.

Once babies were three weeks old (and so long as they were healthy and gaining weight) mums were told to start lengthening the time between night feeds. This didn't mean leaving babies to 'cry it out' for long periods on their own, but teaching them to dissociate waking up and feeding so they didn't automatically expect to be fed if they woke at night. Delaying tactics included changing their nappy, resettling them back to sleep, patting or carrying them. (Note: the current advice from breastfeeding experts is to resist cutting out night feeds for at least the first few weeks of breastfeeding, when feeding on demand is necessary to establish a good and regular supply of milk.)

In fact, many of the women in this study didn't bother with the prolonged evening feed – presumably because they felt uneasy about waking a baby that was already asleep. They also took closer to six weeks to start lengthening the space between night feeds. Still, the fact that there was a difference between the groups suggests that there are measures parents can take to boost night-time sleep.

The use of focal feeds (also known as 'dream-feeds') was investigated in a separate (although small) study of twenty-six women and their newborn babies, half of whom were told to give an extra-large feed between 10 p.m. and midnight and then to gradually increase the amount of time between night feeds by resettling the baby without feeding if they woke at night by swaddling them or changing their nappy. The mothers also tried to settle babies in their cots for daytime naps rather than carrying them and tried to maximize differences between night and daytime (for example, by not using black-out blinds during the day, and trying to engage with babies as little as possible at night).

By the time these babies reached eight weeks, all of those following the programme were sleeping for regular five-hour stretches at night, compared to just 23 per cent of babies that weren't. Even though the babies were feeding less frequently at night, the researchers found that their weight gain and their overall milk intake over twenty-four hours was the same as for babies that fed more often. They seemed to make up for the reduced night feeds by drinking more milk in the early morning.

101

Is co-sleeping good or bad for my baby?

CO-SLEEPING IS THE norm for around 90 per cent of the world's population, yet childcare experts in many Western countries advise against it. Part of the reason is safety, since studies have linked co-sleeping to a slightly increased risk of sudden infant death syndrome (SIDS), although steps can be taken to reduce the risk. However, there's also a prevailing fear that if babies start off sleeping in the parental bed, it will become impossible ever to kick them out of it.

Despite these fears, co-sleeping seems to be on the increase in many Western countries. One survey of US parents found that the number of mothers co-sleeping with their baby for all or part of the night doubled between 1993 and 2000 to 12.8 per cent, while a separate study reported that nearly half of UK mothers share a bed with their baby at some point during the first month of its life; this drops to 29 per cent when babies are three months old.

Those who champion co-sleeping claim that it makes breast-feeding easier, boosts the baby's development and promotes bonding – and that it actually reduces the risk of SIDS, because parents can keep a closer eye on their baby. It is certainly true that studies have found a higher rate of breastfeeding among

mothers who share a bed with their baby compared to b̶
that sleep alone, although it's difficult to disentangle the fact t̶
many women who choose to co-sleep are also particularly keen
breastfeed. One small study that filmed babies either sharing a be̶
with their mums, lying in an open-sided crib that was attached to
the bed or lying in a separate cot found that those in the bed or
crib attempted to feed more often and had more successful feeds
than those in the separate cot, which could help establish a better
milk supply. However, a separate study found no difference in the
rate at which babies gained weight between co-sleepers and babies
that slept in cots.

Evidence that co-sleeping benefits brain development is also
patchy. One study that followed 205 children from birth through
to the age of eighteen found better cognitive performance in areas
such as memory and decision-making among those who co-slept
as babies when they reached the age of six, but the effect was small
and there was no difference in their cognitive, social, emotional
or developmental maturity when they were assessed again aged
eighteen. Meanwhile, a separate study of 175 babies who slept
alone and 29 babies who routinely co-slept found that the solitary
sleepers were generally less irritable than the co-sleepers when
they were assessed at four months of age.

A key problem in all of this is that the definition of co-
sleeping varies between researchers, health professionals and
parents. Researchers often define co-sleeping as sharing the same
room rather than the same bed, but health professionals often
use co-sleeping to describe bed-sharing. Those that champion co-
sleeping often cite a study that found higher self-esteem among
women who co-slept as young children, but it actually looked at
women who spent several years sleeping in their parents' bedroom,
not necessarily sharing a bed with them.

The fact is that very little good-quality research has been done on the physical or emotional benefits of co-sleeping, so the jury is still out.

The issue of safety is also far from straightforward. Those studies that have looked at the characteristics of babies that died from SIDS have identified co-sleeping as a possible contributing factor, but it is difficult to disentangle it from other factors which often go alongside bed-sharing, such as poverty, parents who smoke or the baby sleeping on its front. Depending on the characteristics of individual families, estimates on the risk of SIDS for babies that share a bed with their parents ranges from no increased risk among non-smoking families, to a twelve-fold increased risk for babies sleeping on a sofa with a parent who smokes.

Given this uncertainty, it probably comes down to individual parents to decide what works best for them. If you do want to share a bed with your baby, factors that seem to increase the risk of SIDS include smoking (particularly if the mother smoked during pregnancy, which may affect the development of the brain), sleeping with the baby on a sofa and consuming drugs or alcohol. Overheating and smothering with pillows are also possible risks, so you should take steps to prevent this happening.

102

Will leaving my baby to cry cause any long-term damage?

No SUBJECT IS more divisive among parents than controlled crying. Those who support it claim the pain is only temporary and that teaching babies to settle themselves is one of the greatest gifts a parent could give. Those who oppose it believe that leaving a child to cry is tantamount to cruelty and may cause lasting damage. Parenting books also disagree on the matter. Out of thirty-nine books on infant sleep that are currently available in the US, twenty-four endorse or advocate controlled crying as a means of teaching children to sleep while twelve oppose or warn against it. The remainder take no position.

In terms of teaching babies to sleep, controlled crying does seem to work, although other, less severe methods appear to be equally effective. In 2006, the American Academy of Sleep Medicine (AASM) published a review of fifty-two studies that evaluated the effectiveness of various methods of sleep training, many of which involved some degree of controlled crying.

In its strictest form, controlled crying (also known as 'extinction' or 'crying it out') involves ignoring all crying, tantrums and calls for parents, unless the child is hurt, ill or in danger – and doing this every night, no matter how long the crying lasts. The claim is that if you cave in and respond to the child, it will simply learn to cry for longer the next time. Softer variations include leaving progressively longer intervals before going in and reassuring the child, and then leaving; or staying in the room with the child but ignoring its cries. The ultimate goal is to teach the child to develop its own self-soothing skills so it can fall asleep independently of its parents.

Other strategies for teaching children to sleep involve establishing positive bedtime routines, such as taking a bath and reading them a story, scheduled awakenings, where a child is deliberately roused fifteen to thirty minutes before it would usually wake up and then settled back down while still semi-conscious, and educating parents about establishing regular nap-times, consistent bedtime routines and putting babies into their cots when they're drowsy, but not asleep.

According to the AASM review, all of these strategies worked to some degree and, of the few studies that conducted head-to-head comparisons of the different methods, there was little evidence to suggest that any of them was vastly superior in terms of effectiveness in the long run. Strict controlled crying seemed to work faster than scheduled awakenings in babies that often woke at night, but the different forms of controlled crying seemed to be pretty much on a level playing field in terms of teaching babies to sleep, along with establishing positive bedtime routines.

Parents should also consider a baby's age before considering any form of sleep training.

Babies younger than three months

At least two randomized controlled trials have found that trying to use any form of sleep training on babies younger than three months is ineffective, and doesn't decrease crying. There are steps you can take to prevent bad habits from forming, however (see 100: 'How can I get my baby to sleep through the night?').

Babies aged three to six months

Most experts agree that controlled crying shouldn't be used on babies under six months of age, as they may still need to feed during the night. At this age, you might try working on establishing positive bedtime routines and making sure your baby gets regular naps during the day – or try scheduled awakenings if your baby often wakes during the night.

Babies older than six months

If you and your baby still aren't getting enough sleep and you're considering controlled crying, you can take comfort from several studies that have investigated whether it has any long-term adverse effects. Their general conclusion is that, overall, sleep training makes babies more secure, predictable and less irritable than they were before treatment. An Australian study of 225 babies whose parents used controlled crying at eight to ten months of age also found that controlled crying reduced babies' sleep problems by around 30 per cent within four months, although 15 per cent of babies didn't respond to it. When parents completed questionnaires assessing their children's emotional and behavioural development at various points until the age of six, there seemed to be no difference between children who had received sleep training as infants and those who hadn't.

However, there is one final note of caution. In 2012, a study was published suggesting that controlled crying did have a short-term physical effect on a baby's stress levels even after it had stopped crying out for its parents at night. Wendy Middlemiss and her colleagues at the University of North Texas measured daily levels of the stress hormone cortisol in twenty-four babies aged four to ten months as they underwent a five-day residential sleep-training programme at a hospital in New Zealand. The programme involved a typical regime of controlled crying, in which mother and baby were separated at bedtime and a nurse settled the baby into its cot. It was then left to fall asleep by itself – although a nurse would go in and check on the baby every ten minutes if it continued to cry. The mother was not allowed to comfort the baby, but was in a nearby room and could hear its cries.

Levels of cortisol were similar in mums and babies before sleep training commenced and, perhaps unsurprisingly, they rose considerably after the first night. The same thing happened on the second night, but by the third night the babies had learned to settle themselves to sleep and cried considerably less. At this point, the mothers' cortisol levels began to fall, and they also expressed relief that their babies were beginning to respond to sleep training and they were finally getting a decent night's sleep. However, measurements of cortisol revealed a disconnection between the babies' outwardly calm appearance and what was going on inside. Although the babies no longer cried, they continued to experience high levels of cortisol, similar to those on the first night. This might suggest that, although they're calm on the outside, they are still unhappy on the inside.

No further measures of cortisol were taken after the programme ended, so it's impossible to know at this stage whether the babies' stress levels eventually normalized – although the fact that other studies have found no lasting negative effects from controlled crying is somewhat reassuring. Further studies are clearly needed to confirm how long babies remain stressed. 'It may well be that the infant is still in a transition period and will be fine,' says Middlemiss, 'but the disturbing thing for me is that the mother can't choose whether or not to respond to the infant's distress, because she can't see it.'

Given that less severe forms of controlled crying seem to be just as effective as the strict version in terms of curbing sleep problems, I'd personally be tempted to try one of these variations first.

103

What's more effective: introducing strict routines or immediately responding to my baby's demands?

As WITH CONTROLLED crying, parents tend to split over whether babies should adhere to strict schedules or have Mum and Dad at their beck and call. In *The New Contented Little Baby Book*, Gina Ford suggests that in order to get your baby to sleep through the night from an early age you need to establish the right associations and structure your baby's feeds and naps from the day you arrive home from hospital. At the opposite end of the spectrum, attachment parenting gurus such as Jean Liedloff advocate prolonged holding, frequent breastfeeding, co-sleeping and rapidly responding to a baby's cries in order to foster strong emotional bonds between mother and baby.

In a recent study, Ian St James-Roberts and colleagues at the Thomas Coram Research Unit in London compared the impact of these very different parenting styles on the amount of time young babies spent crying and sleeping. They recruited three groups of parents: one from London, one from Copenhagen and a third group of mixed-nationality parents who practised an extreme form of attachment parenting (also known as proximal care), in which babies were carried around for much of the day, breastfed on demand, and parents rapidly responded to their baby's cries. Parents were asked to keep daily diaries of what they did and how their babies responded to them at various points during their baby's first twelve weeks. They also completed questionnaires about infant feeding and sleeping patterns.

Parents in the proximal care group spent around sixteen and a half hours a day holding their ten-day-old babies and fed them fourteen times a day, while London parents tended more towards the tough-love approach, holding babies only for around half this amount of time and spacing out feeds so they ate just eleven times a day. Babies in Copenhagen fell somewhere in between – held for nearly ten hours a day and fed twelve times. London parents also left their baby to cry for around three times longer each day than either the Copenhagen or proximal-care parents.

At twelve weeks of age, 85 per cent of proximal-care and 70 per cent of Copenhagen babies were still exclusively breastfed, compared to just 37 per cent of London babies (though this could be for cultural reasons), and most proximal-care parents co-slept with their infants, while just 16 per cent of Copenhagen parents and 9 per cent of London parents did the same.

So how did the babies respond? Those living in London spent around 50 per cent more time fussing and crying than those in the other groups, although there was no difference in bouts of inconsolable crying (see 89: 'What causes colic?'). When parents were asked how many nights in the past week their twelve-week-old baby had slept for five hours or more, London and Copenhagen parents reported around five such nights, while proximal-care parents reported just three and a half nights on average.

Although this is just one study, it does suggest that the extreme form of proximal care is more likely to result in frazzled parents, without any obvious benefit in terms of sleep quality or the amount of time babies spend crying, compared to the more moderate approach of the Danes. By the parents quickly responding to their baby's cries during the daytime, the Danish babies spent less time crying overall (although we don't know if extra crying is actually harmful to babies), and they seemed to sleep just as well. On the other hand, the tough-love approach of the London parents didn't mean they got any less sleep – they may have just had to put up with sore eardrums during the day.

The White Stuff

104

How does breastfeeding work?

It's one of the most natural processes in the world, yet until very recently the mechanism by which babies manage to extract milk from the breast was unclear. Some thought babies squeezed milk out of the nipples by kneading them in their mouths – much as a milkmaid squeezes milk from a cow's udder – while others proposed that the milk was removed by suction.

In 2010, Australian researchers finally solved the mystery after combining ultrasound imaging of babies suckling on the breast with measurements of the vacuum created by their mouths. 'It's not a milking action at all,' says Donna Geddes of the University of Western Australia in Crawley, who led the research. 'What we see is that when the tongue is lowered and the vacuum is applied, that's when the milk is coming out of the breast.'

They also found that babies who struggle to breastfeed generate far weaker vacuums than those who breastfeed successfully, which may help explain why premature babies often struggle to feed: their mouth muscles aren't yet strong enough to create a vacuum.

The study also emphasizes the point that painful crushing or distortion of the nipple isn't necessary for breastfeeding to occur. If your nipples are coming out of your baby's mouth distorted, this probably means that your baby isn't latching on correctly, and you should either try repositioning them or speak to a lactation consultant.

105

Should I feed on demand or wait three to four hours between feeds?

IT'S PRETTY MUCH undisputed that if you plan to breastfeed, you should try to feed whenever your baby appears hungry, for at least the first couple of weeks. The amount of milk that women produce six days after birth is strongly linked to the volume they will produce at six weeks, so it seems that frequent stimulation of the breasts during these very early days really is needed to get the breasts producing enough milk in the first place. It's also true that if you start substituting breastfeeds with formula milk, then your milk production will start to fall.

But what about once breastfeeding has been successfully established? Women are usually told to 'feed on demand', without being told what this actually means. Does it mean feed whenever the baby cries? Or only when the baby is hungry, in which case, how do you know if the baby is hungry rather than tired or uncomfortable? The simple answer is that there is no hard-and-fast rule. Studies have shown that individual babies show a three-fold variation in the amount of milk they consume per day, as well as a wide variation in their patterns of milk intake. By the age of one month, the average baby drinks 750–800ml of milk per day, but a recent Australian study found that, although babies under six months fed eight times every twenty-four hours on average, some took as few as four feeds and others as many as thirteen feeds during this time.

Having got the hang of breastfeeding, many women want to try to space out breastfeeds – particularly at night. When this study compared the amount of breast milk consumed by babies that did and didn't feed during the night, it found no significant difference in the total amount of milk they consumed over twenty-four hours. The main difference was that the night-feeders drank the biggest quantities of milk at night, while those that skipped night feeds drank more in the morning to compensate. This should serve to reassure women that if their baby doesn't feed during the night, they will probably still be getting enough to eat overall.

Other common worries include whether babies who take frequent small breastfeeds are getting enough high-fat milk to fill them up, and whether there is enough milk in the breasts to satisfy a baby's appetite. Once breastfeeding is established, the average breast holds 179ml (or 179g) of milk – or almost twice the amount most babies will consume at a single sitting. However, the storage capacity of individual women ranges from 74ml to 382ml per breast. In general, women with bigger storage capacities give feeds less often, but not always. So you shouldn't simply assume that your baby is asking to feed more often because you're not making enough milk to quench its thirst. Neither does there seem to be much relationship between the size of your breasts before pregnancy – or how much the breasts grow during pregnancy – and how much milk they can store. Indeed, studies have found that the amount of breast growth during pregnancy is relatively consistent between mothers, regardless of their original breast size. 'Of course, if the breasts are very small they will not be able to store a large amount of milk, but apart from that, there will be no relationship between bra size and breast-milk storage capacity,' says Jacqueline Kent of the University of Western Australia, who researches how breastfeeding works. One way of estimating how much milk your breasts contain is to use a breast pump to empty them before you would usually feed your baby, and then give the milk to the baby in a bottle, although this is only a rough guide, as pumps don't necessarily extract as much milk as babies do – particularly if, like me, you find them uncomfortable to use for more than around ten minutes at a time, when your baby would feed for longer.

In general, the fat content of breast milk changes as babies progress through a feed, starting with more watery 'fore-milk' and progressing to richer 'hind-milk', but how much it changes varies according to how often women give feeds. Studies show that when babies take six to nine large breastfeeds per day the fat content changes from around 4.3 per cent at the start of the feed to 10.7 per cent by the end, but the fore-milk tends to be richer in women who give feeds more frequently. When babies take fourteen to eighteen small breastfeeds each day, the fore-milk contains around 4.8 per cent fat and the hind-milk around 8.2 per cent fat. Regardless of how often they feed, there seems to be little difference in the total amount of fat that babies receive.

That said, some common breastfeeding advice – such as only allowing the baby to feed for ten minutes per breast and not allowing 'comfort suckling', which usually occurs towards the end of a feed and can be felt as rapid flutter-like sucks and periodic swallows – may reduce a baby's chances of getting their fill of creamy hind-milk.

If you're uncertain about whether your baby has fully drained the first breast and should be offered the second, the best measure is whether or not you can still hear them swallowing, says Pamela Douglas, a breastfeeding expert at the University of Queensland in Australia. If they have stopped swallowing, it probably means they have finished feeding, but you should try offering the second breast to make sure.

106
Is breast really best for babies?

THE MANTRA THAT breast is best probably is true, at least for the first few months of a baby's life. The most convincing evidence for the superiority of breast milk over formula relates to the prevention of stomach and respiratory infections, including colds, flu and diarrhoea, among breastfed babies. This is particularly important in developing countries, where it may be harder to sterilize bottles, find clean water to prepare formula, and where diarrhoea or lung infections may quickly become life-threatening as a result of inadequate medical care. It was largely for these reasons that the World Health Organization recommended that women should exclusively breastfeed for six months before starting to introduce solid foods, but ideally continue to breastfeed their baby until the age of two.

Even in developed countries, it is estimated that 53 per cent of hospitalizations for diarrhoea among babies could be prevented each month if everyone breastfed their babies exclusively, and breast-feeding has been shown to cut the risk of necrotizing enterocolitis (a serious gastrointestinal disease that mostly affects preterm babies) by around 55 per cent. It reduces the risk of hospitalizations for peneumonia by around 72 per cent. Meanwhile, there's pretty good evidence that breastfeeding cuts the risk of SIDS by around 36 per cent – even though no one really understands what causes it.

Mild diarrhoea aside, all of these illnesses are relatively rare in developed countries, so if you can't breastfeed for whatever reason, it's unlikely that your baby will come to harm through drinking formula. For example, in the case of SIDS, it is estimated that 5,500 children would have to be breastfed to prevent one death. And in the case of pneumonia, twenty-six would have to be breastfed for four or more months to prevent one hospitalization. Breastfeeding is no guarantee against illness – it merely reduces the risk – and the protection only lasts for as long as the woman is breastfeeding and ends soon after she stops.

Of course, the majority of women stop breastfeeding long before the two years recommended by the WHO, often for practical reasons such as having to return to work. Many women struggle to breastfeed in the first place, and it is often with an extremely guilty heart that they resort to the bottle.

Are they right to feel guilty? Besides reducing the risk of infections in the short term, there are plenty of claims for additional benefits of breastfeeding. Many health organizations, including the UK's National Health Service and the American Academy of Pediatrics, state that breastfeeding also offers long-term protection against obesity, eczema and diabetes. Others, such as the website Ask Dr Sears, state that breastfeeding boosts IQ, eyesight, response to vaccination, and reduces the risk of childhood cancer or the need for orthodontics.

However, once you start delving into the medical literature it becomes clear that the evidence for breast being overwhelmingly better than bottle-feeding on these longer-term health measures is less clear-cut. For example, studies have shown that babies who are not breastfed have 1.3–1.9 times as many allergy-related problems, are two to four times more likely to get (admittedly rare) childhood cancers, and are 1.2–1.6 times as likely to be overweight. A key problem is that many of these studies asked women to recall their breastfeeding habits from years ago, rather than monitoring children as they grow up, and their memories may have become clouded by what has subsequently happened to their child. When it comes to things like obesity, IQ or high blood pressure, it is also difficult to disentangle socio-economic factors, because mothers who breastfeed often come from different backgrounds to those who formula-feed.

The best way to take all of these factors into account would be to do a randomized controlled trial (RCT), in which similar women would be told either to breastfeed or formula-feed for a fixed period of time, and the health of their children would then be closely monitored for years afterwards. No such trial has been done. For one thing, you can't tell women what to feed their babies. For another, many women who start off breastfeeding give up after a few weeks or months, or do complicated things such as breastfeeding during the daytime and introducing a bottle of formula at night.

Perhaps the closest anyone has come to this sort of RCT was a study of 16,491 women from Belarus conducted by Michael Kramer of McGill University in Montreal, Canada. In this study, half of the women were given advice and support regarding breastfeeding, while the remainder received no such help. Providing this help worked to some extent – 43 per cent of women who received advice and support were exclusively breastfeeding when their babies were three months old, compared to just 6 per cent in the control group. Like other research, it found that exclusive breastfeeding reduced the risk of gastrointestinal infection and eczema during infancy. However, when Kramer's team re-examined the children aged six and a half, they found no difference in height or obesity, no reduction in blood pressure, allergy, asthma, or tooth decay, and no difference in behaviour between the two groups.

These findings have since been backed up by a large study in the UK, which has been following the progress of more than 14,500 women and their babies since the early 1990s. Although it found that breastfeeding was associated with less obesity and lower blood pressure, it was impossible to disentangle this from the fact that women who breastfeed in the UK are more likely to come from middle- or upper-class backgrounds. When the same researchers looked at babies raised in Brazil, where breastfeeding isn't linked to social status, the links to obesity and blood pressure disappeared.

That's not to say that breastfeeding doesn't have long-term benefits, it's just that a lot more work needs to be done before we have a definitive answer. 'In terms of a long-term effect on metabolism or the immune system, the evidence is either meagre in terms of the number of studies, or conflicting because there are a lot of studies that don't agree,' says Kramer. It's also likely that any protective effect will be small compared to more well-established risk factors, such as having a strong family history of obesity or disease.

However, there was one area where both Kramer and the Avon Longitudinal Study found strong evidence of a benefit: brain development. In Kramer's study, the breastfed children scored around eight points higher on IQ tests when they were six and a half than those who were bottle-fed, and, unlike many previous studies, the researchers had taken the mother's intelligence and socio-economic background into account, so this couldn't explain the difference. 'It's not really a delayed effect, because we know that the brain is largely developing in the uterus and during the first year or two of life,' says Kramer. As for why breastfeeding should boost brain development, no one really knows. We don't even know if it's anything in the breast milk. 'It could be a result of the physical contact between the mother and baby, or even increased verbal contact between the mother and baby during breastfeeding,' Kramer says.

107

How long do I need to breastfeed in order for my baby to reap the benefits?

ALL MUMS WANT to do the best by their babies, but suggesting that they have to breastfeed for six months (let alone two years) or they will be failing their child is not only unrealistic for most working mothers, it may not be supported by evidence regarding health benefits. 'The strongest evidence for immediate breastfeeding benefits is exclusive breastfeeding for the first three to four months of life,' says Ronald Kleinman, physician-in-chief at Massachusetts General Hospital for Children. 'This impacts resistance to upper respiratory infections and diarrhoea, and potentially improves responses to vaccines – although the evidence for this isn't as strong.'

Studies that have found a reduction in the risk of SIDS also followed babies for only six months, and the majority of SIDS deaths occur between the ages of two and four months.

As for longer-term benefits, such as protection against obesity or diabetes, the existing evidence also points to breastfeeding during the earliest months being key. For protection against allergic disease and type 1 diabetes, the mothers in the studies breastfed for at least three to four months, while in the studies suggesting some degree of protection against childhood cancer, babies were breastfed for six months.

After this, it's hard to know if additional breastfeeding offers significant long-term benefits for babies besides the immediate protection it provides against infectious diseases, which, although important in developing countries, is arguably less of an issue in rich countries with adequate medical care. That's not to say that extended breastfeeding isn't beneficial, just that it has barely been studied – in part because so few women in Western countries do it. 'There is really no solid evidence that introducing some formula at six months, or weaning from the breast and changing to formula at that time, has any health implications,' Kleinman says.

One possible exception is brain development, where studies have indicated that babies get more benefit the longer they are breastfed. 'At least in developing countries, there's evidence that breastfeeding beyond a year, or at least up to a year, is important,' says Michael Kramer.

However, he emphasizes that women who don't breastfeed for this long should not necessarily feel guilty about it. 'There are many other issues that are as important, if not probably more important, than breastfeeding,' he says. For example, sitting your child in front of a TV and not playing with it is likely to have a bigger impact on its IQ than not breastfeeding. 'In general, a happy mother is a better mother,' says Kramer. 'You're not guaranteed to be a good mother just because you follow the recommendations and breastfeed for two years.'

Finally, women are often told that breastfeeding provides

numerous benefits for their own health, such as protection against certain types of cancer, diabetes or obesity (see 74: 'Do women who breastfeed really lose their baby-weight faster than those who bottle-feed?'). It's true that some studies have found health benefits, and the effect here seems to be cumulative – so it's the total amount of time you spend breastfeeding all of your babies that matters. For type 2 diabetes, there is a decreased risk of approximately 4 to 12 per cent for every year of breastfeeding, while for breast and ovarian cancer, breastfeeding for more than twelve months is associated with a 28 per cent decrease in cancer risk.

108

Could combining breast- and formula-feeding offer babies the best of both worlds?

WHETHER IT'S THE occasional bottle of formula if you're going out for the evening or a regular bottle to help your baby to sleep through the night, plenty of parents decide to combine breast-feeding with bottle-feeding. Some Latino communities even have a phrase for it: '*los dos*'. Many believe it offers the best of both worlds, combining the protective effects of breastfeeding with the additional nutrients of formula milk.

Despite its prevalence, such combination-feeding has barely been studied. The World Health Organization doesn't even distinguish it from the introduction of solid foods, despite the fact that the baby is still getting all of its nutrients from milk. From their point of view, you either breastfeed exclusively, or it doesn't count.

But can you really have the best of both worlds? Those studies that have been done suggest that while combination-feeding does provide some of the health benefits of breastfeeding, these are reduced. Take the common ear infection otitis media. Exclusive breastfeeding for more than three months roughly halves the rate of ear infections, while in babies fed a combination of breast and formula milk the risk is cut by around 23 per cent. Exclusive breastfeeding also has roughly double the protective effect against sudden infant death syndrome (SIDS) compared to breastfeeding for some of the time. There's also some early evidence that breast milk produced in the evening may contain chemicals that help babies sleep, so if you do use formula for a night-time 'dream feed', you may miss out on this benefit (see 113: 'Is morning breast milk any different to evening breast milk?').

What's not so clear is whether introducing the occasional bottle of formula milk makes it any harder to maintain breastfeeding in the long term – a common concern of health professionals, because we know that frequent feeding in the early days is needed to establish a good milk supply. One study that compared babies fed a combination of breast- and formula milk during the first week of life confirmed that the mums of combination-fed babies were less likely to continue breastfeeding, but only if they were white. Black and Hispanic mothers, on the other hand, seemed quite capable of continuing to combine breast- and bottle-feeding and were no less likely to have stopped breastfeeding by the time their babies reached three months of age.

This difference implies that the decision to switch from combination- to formula-feeding is a cultural one, rather than the babies preferring the flavour of formula milk or refusing to take the breast after sucking on an artificial teat (see 110: 'Does "nipple confusion" really exist?'). In other words, if you have the willpower to combine breastfeeding with formula-feeding, there doesn't seem to be a biological reason why you can't do it. Meanwhile, plenty of women who have started off combination-feeding, perhaps for medical reasons while in hospital, have managed to switch to exclusive breastfeeding afterwards – it just takes extra effort to get the milk supply going.

One final word of warning: most experts agree that regular breastfeeding is essential in order to maintain the milk supply. So although giving babies the odd bottle of formula seems unlikely to make much difference to either their health or your ability to continue breastfeeding, regularly skipping breastfeeds will cause the milk supply to diminish.

109

What is in formula milk, and why does it taste fishy?

ALL BASIC INFANT formula is made from cows' milk, although the proportions of the two main milk proteins, casein and whey, have been adjusted to mimic that of human milk more closely. Surprisingly, there's no difference between formula and breast milk in terms of the number of calories they provide, with both containing around 67kcal per 100ml of milk.

Many manufacturers produce formulas targeted at different age ranges. In 'first milk' formulas, relatively more whey protein has been added, which makes it easier to digest. Formula marketed at 'hungry babies' contains less whey and more casein, making it closer in composition to regular cows' milk. The idea is that casein forms a solid curd in the stomach and therefore takes longer to digest – which means, theoretically, the stomach stays fuller for longer – but it contains no more calories than normal formula and usually contains less fat. Neither is there any convincing evidence that babies are any less hungry on it, although they may be more likely to get constipated.

The same is true of 'night-time' milks, which contain starch to thicken them. The UK's Scientific Advisory Committee on Nutrition (SACN) recently published a statement saying that there is no evidence that these offer any advantage over normal formula in terms of settling babies at night.

Meanwhile, follow-on formula is usually marketed at babies aged six months and up, and contains more iron, vitamin C, vitamin E, zinc and calcium than normal formula. However, most babies are eating at least some solid food by this stage so will be getting additional nutrients from that. Iron is also more readily absorbed from solid food than formula. The SACN has advised that there is no need to use follow-on milk instead of normal formula or cows' milk once babies reach one year of age. The American Academy of Pediatrics also says that follow-on formulas offer no clear advantage over infant formulas designed to meet all nutritional needs over the first year, so you may as well let price guide you.

There's also little difference between brands of basic formula in terms of their nutritional content. Within Europe and the USA at least, the use of the term 'infant formula' is regulated, and all must meet minimum nutritional requirements. All are fortified with iron, which may be what gives formula a slightly metallic or fishy taste.

However, some companies have started adding extra ingredients to basic formula, often alongside claims that these will boost babies' health:

Long-chain polyunsaturated fatty acids

Fatty acids such as docosahexaenoic acid (DHA) and arachidonic acid (AA) are sometimes added, with the claim that these promote eye and brain development. Several recent studies have supported these claims – at least when higher doses of DHA and AA are added (so each comprises at least 0.3 per cent of total fatty acids) – although others have found no benefit. A review by the Cochrane Collaboration showed there was no clear and consistent benefit of adding fatty acids to formula in terms of eye or brain development, or physical growth in healthy babies that weren't born prematurely.

Prebiotics and oligosaccharides

Another recent addition to formula milk is so-called 'prebiotics', or oligosaccharides (such as GOS or FOS). These are complex sugars found in breast milk which are thought to promote the growth of certain bacteria in the gut. A recent review concluded that adding them did increase the number of 'friendly' bifidobacteria in the gut, and babies had softer and more frequent stools. However, a Cochrane review found no convincing evidence that prebiotics helped prevent allergies or food hypersensitivity, as some have claimed.

Probiotics

These are live bacteria which are sometimes added to formula milk. There is fairly strong evidence that probiotic supplements can reduce the development of eczema in babies with a strong family history of it.

Nucleotides

These form the basis of the DNA in every cell and are added to most formula these days. There is some evidence that these may reduce the incidence of diarrhoea and boost response to vaccines after immunizations, although many of these studies were funded by formula manufacturers, and other studies have found no effect. Given that most formulas contain nucleotides, and the jury is still out on whether they have much benefit, it's probably not something to worry about.

Lactose-free or hypoallergenic formulas

Doctors sometimes recommend that babies with suspected feeding intolerances such as lactase deficiency or milk-protein allergy be fed a lactose-free or hypoallergenic formula. Paediatricians have voiced concerns that such conditions are being over-diagnosed and point out that spitting up milk is fairly normal, and not necessarily an indication of an allergy or intolerance. These specialized formulas contain carbohydrates from corn or sucrose, rather than the carbohydrate found in milk called lactose. Hypoallergenic formula has also been treated with enzymes to break down the casein protein, against which some babies develop an immune reaction.

Soy formula

This is popular among parents who think it will be more digestible and less likely to trigger an allergic reaction, but a Cochrane review found that there wasn't enough evidence to recommend soy formula for prevention of allergy or food intolerance in babies at high risk of these conditions. Concerns have also been raised about phytoestrogens in soy milk mimicking the female hormone oestrogen. The European Society for Paediatric Gastroenterology, Hepatology and Nutrition (ESPGHAN) recommends that soy formula only be used in babies older than six months who have been formally diagnosed with an allergy to milk protein.

110

Does 'nipple confusion' really exist?

PARENTS ARE OFTEN told to avoid giving their baby a bottle or dummy during the first three to four weeks of life as it will cause 'nipple confusion'. The idea is that the sucking action of breastfeeding differs from the technique used to extract milk from a bottle, so babies may become confused about how to breastfeed and even reject the breast as a result. Some claim that just one bottle-feed can cause nipple confusion and that this might ultimately result in failure to establish breastfeeding. Yet many women physically can't breastfeed their newborn baby in the immediate aftermath of birth, so milk has to be given some other way. Traditionally, hospitals used bottles containing either expressed breast milk or formula, but fears about nipple confusion mean that many currently prefer to use a cup or feeding tube instead.

Although it's generally true that mums who start off mixing breast- and bottle-milk are less likely to continue breastfeeding, the silicone teat isn't necessarily to blame. Alison Holmes and her colleagues at Dartmouth Medical School in New Hampshire studied 802 mother–baby pairs in which the baby had a combination of breast- and bottle-feeding during the first week in hospital. Although such combination-feeding shortened the amount of time that white babies were breastfed, this was not the case in Latino or black babies. While it's possible that black and Hispanic babies are biologically different to white babies, this seems improbable. More likely is that the use of a silicone teat was irrelevant, and the white women were giving up breastfeeding for different reasons, possibly because they expected breastfeeding to be a struggle after bottle-feeding and this then became a self-fulfilling prophecy. By contrast, Hispanic women have a strong tradition of combining breast- and bottle-feeding and see it as completely normal (see 108: 'Could combining breast- and formula-feeding offer babies the best of both worlds?').

In a separate study, the Cochrane Collaboration reviewed four trials comparing cup- and bottle-feeding of breast milk to newborn babies. Although three of the studies found that the cup-fed infants were slightly more likely to be exclusively breastfed when they were discharged from hospital, there was no difference in the rate of exclusive breastfeeding when the babies were compared at the ages of three and six months. The results of one of the studies also suggested that cup-fed infants spent an average ten days longer in hospital than bottle-fed infants.

Yet more evidence against the existence of nipple confusion comes from studies of the use of dummies by young babies. Although some studies have found a link between dummy use and mothers giving up on breastfeeding, these couldn't show that the dummies themselves were to blame. What's more, several studies that combined the results of multiple clinical trials investigating the link between breastfeeding and dummy use found that, when used in healthy breastfeeding babies, dummies had no significant effect on the proportion that were still breastfeeding at three months (see 97: 'Are dummies good or bad for babies?').

111

Does what I eat change the flavour of my milk?

YOU MIGHT ASSUME that an exclusive diet of milk would get rather boring after a while, but human milk provides a cornucopia of different flavour combinations which might even influence a baby's food preferences in later life. One of the most famous examples of this is carrots, as one study found that children whose mothers drank carrot juice during pregnancy or while breastfeeding showed a preference for carrot-flavoured cereal (see 7: 'Can Mum's food fads influence her baby's palate?').

In a separate study, breastfeeding mums were given edible capsules containing the same compounds that give caraway seed, menthol, banana and liquorice their flavours. Although these different flavours took varying amounts of time to show up in breast milk, appear they did. After just one hour, the milk took on the flavour of bananas, while after two hours it began to taste of caraway and liquorice. Menthol also got into the milk after about two hours and then stayed there for a further six hours. All four flavours had disappeared eight hours after the capsules had been eaten. However, there was considerable variation between the women in how the flavours were transferred to milk. If you suspect that something you've eaten may be upsetting your baby's delicate palate, eight hours should be long enough for any unpleasant flavours to be eliminated, says Helene Hausner at the University of Copenhagen in Denmark, who led the study.

Hausner has also found that different brands of formula milk vary in taste, so if your baby doesn't seem to like one brand, it may be worth trying another – although different brands generally have the same basic nutritional content.

112

How much alcohol gets into breast milk?

YOU'VE GONE THROUGH nine months of abstinence during pregnancy and are just looking forward to enjoying wine with dinner again when a well-meaning friend asks, 'Should you really be drinking if you're breastfeeding?'

However, one midwife assured me that drinking actually aided breastfeeding, by making you more relaxed and stimulating the milk let-down reflex. Who should you believe?

The fact is that alcohol does get into breast milk in about the same quantities as you would find in the blood, but it isn't stored there. This means that so long as you wait to clear the alcohol from your system before breastfeeding, then your baby won't be exposed to it at all.

On a practical level, this means that if you were to drink an entire bottle of wine in the evening (about nine units, although wines vary in alcohol content), it would take around nine hours for this alcohol to leave your body, which means that you could probably breastfeed the following morning without having to worry about it. And because alcohol isn't stored in the breasts there's no need to pump out the residual milk and pour it down the drain.

It's also a relief to know that a toxic breakdown product of alcohol called acetaldehyde, which is partly responsible for hangover symptoms such as headaches and vomiting, doesn't pass into breast milk.

However, even if you decide to breastfeed immediately after having a drink, the alcohol that your baby received would be greatly diluted as a result of passing through your body. In the 1980s, researchers from New Zealand encouraged eight nursing mothers to drink as much alcohol as they could manage in as short a time as possible and then measured how much of it they could detect in their milk. They found that the alcohol level in blood and milk peaked between one and a half to two hours after drinking and then began to fall, but even at their peak, the amounts were far less than would have been present in the original drink. For example, the researchers calculated that if a six-month-old baby weighing six and a half kilograms drank 180ml of milk when its mother's alcohol level was 119mg/100ml blood (the legal UK drink-drive limit is 80mg/100ml), the baby's blood alcohol level would rise to just 6mg of alcohol/100ml. Although babies have less of the enzyme that breaks down alcohol than adults, 'It is improbable that occasional exposure to alcohol of that quantity would affect the child,' writes Margaret Lawton of the Department of Scientific and Industrial Research in Auckland, who led the research.

Although there's currently no evidence to suggest that such tiny amounts of alcohol cause any lasting damage to babies, it may still have subtle effects on their behaviour. For example, Julie Mennella of the Monell Chemical Senses Center in Philadelphia has been studying the transfer of alcohol into breast milk for more than two decades, and has found that just one to two alcoholic drinks will change the flavour of breast milk. Babies are thought to learn about flavours through drinking breast milk, so it is possible that regular drinking may have an influence on their taste for alcohol in the future. In the short term at least, Mennella has found that the babies of mums who regularly enjoy a drink while breastfeeding seem to be more interested in chewing a toy that smells of alcohol than if their mother abstains.

Consuming alcoholic breast milk may also affect the quality of a baby's sleep. In a small study, Mennella gave babies 100ml of breast milk flavoured with 32mg alcohol – the sort of amount they'd receive if their mothers had been drinking heavily. She found that the babies tended to fall asleep faster but slept for significantly shorter periods of time in the three and a half hours after drinking the alcoholic milk compared with when they drank pure breast milk. 'Babies have less active sleep and then compensate for it in the next few days if Mum doesn't drink again,' says Mennella. She adds that similar changes in sleep patterns have been seen in older foetuses whose mothers drink during the latter stages of pregnancy.

Finally, Mennella has also blown the lid on the received wisdom that a glass of wine helps to stimulate the milk let-down reflex. In fact, alcohol blocks the release of oxytocin, leading to an overall reduction in milk production. Mennella found that babies consumed 20 per cent less breast milk on average in the three to four hours after their mothers had been drinking, compared with babies whose mothers hadn't had an alcoholic drink. What's more, they seemed to increase the intensity of their sucking and the number of times they demanded a breastfeed in the next eight to twelve hours to compensate – not necessarily what you want when you're suffering from a hangover. The mothers who drank also noticed that their breasts felt fuller, which may be where the myth that alcohol increases milk production sprang from.

113

Is morning breast milk any different to evening breast milk?

IN 2009, SPANISH RESEARCHERS made a fascinating discovery: breast milk produced during the evening appears to contain natural sedatives which may promote drowsiness. These chemicals (called nucleotides) play other important roles within cells, but several have also been implicated in the processes of sleep.

Cristina Sánchez at the University of Extremadura in Badajoz, Spain, looked at levels of the three nucleotides most strongly linked to sleep in the milk of thirty breastfeeding women over the course of twenty-four hours. She found that levels of a nucleotide called 5'AMP were highest in the evening, while levels of 5'GMP and 5'UMP increased as the night wore on. All were found at far lower concentrations during the daytime.

The most likely explanation is that 5'AMP in breast milk stimulates the release of another sleep-promoting chemical in the brain called GABA, while 5'GMP fuels the production of melatonin, which helps to regulate the body's natural clock. 5'UMP is known to promote both REM and non-REM sleep. These are early findings, but they at least hint that babies given a night-time feed of formula or breast milk expressed during the day won't be getting this added natural sleep-aid.

Night feeds may have some added sleep benefits for mums too. Breastfeeding activates the release of a hormone called prolactin, which promotes drowsiness and also increases the relative propor-tion of deep sleep that you have. This is the kind of sleep that refreshes you and leaves you ready to tackle the next day. A small study found that women who breastfed their babies spent more time in deep sleep and less in light sleep compared to those who bottle-fed their infants, even though they spent a similar amount of time awake. One implication of this might be that night feeds have less of an impact on women who breastfeed because they get more good-quality deep sleep to compensate.

114

Can men lactate?

MANY FATHERS ENVY the closeness that breastfeeding brings, while I'm sure many mums wish at certain points that their partner could share the job of feeding.

Men actually possess all of the equipment needed to breast-feed, including milk ducts and mammary tissue; they just don't produce enough of the hormones to trigger milk production under normal circumstances. One of the main hormones involved in breastfeeding is prolactin, which new fathers are known to produce in small quantities and which seems to stimulate paternal behaviour (see 32: 'Do men change when they become dads?').

Prolactin is produced by the pituitary gland, and men who develop pituitary tumours sometimes lactate. Certain drugs, including the anti-psychotic drug thorazine and the heart drug digoxin, can also stimulate milk production in men, as can a medical condition called galactorrhea, in which the nipples begin to leak a milky discharge. Galactorrhea is associated with high levels of prolactin and low levels of testosterone.

Intriguingly, this 'male milk' seems to be similar in composition to that produced by women. In 1981, researchers at the University of Western Australia tested the nipple secretions of a

27-year-old man with galactorrhea and found in it comparable levels of protein, sugar and electrolytes to female breast milk and colostrum.

Unfortunately, I've failed in my attempts to convince my husband that he should try pumping his breasts or taking hormone supplements to see if any milk comes out. If anyone else succeeds, please let me know!

Weaning

115

When should I wean my baby on to solid foods?

JUST A DECADE ago, parents were told to start introducing solid foods from four months of age. Then, in 2002, the World Health Organization brought out guidelines saying that babies should be exclusively breastfed for six months and solids could be introduced after that. The main reason given was that breastfeeding provides some protection against infection, and the more breast milk babies receive, the greater the level of protection (see 106: 'Is breast really best for babies?').

Two months is a long time in the life of a baby, and this change has left some parents in a state of confusion. Does it mean that introducing solids earlier than six months is bad for babies? And what should parents do if their baby is formula-fed?

Certainly, the consensus among paediatricians is that starting solids before the age of four months is a bad idea. Different studies have shown an increase in symptoms of asthma, eczema and obesity among children given solids before four months.

However, the evidence is by no means clear-cut: a separate review of five studies conducted in the UK found no difference in the health of babies given solids at the even younger age of twelve weeks.

Once babies reach four months, the picture gets even fuzzier. Both the American Academy of Pediatrics and the European Society for Paediatric Gastroenterology, Hepatology and Nutrition say that solid foods can be introduced between four and six months of age – earlier than the WHO recommendation of six months.

One reason parents might want to start introducing solids before six months is because they think that their baby needs the extra nutrients, or that it will help them sleep longer at night. Although the UK review found that breastfed babies were more likely to be sleeping through the night if given solids at twelve weeks of age, other studies have found no difference in sleep patterns between early- and late-weaned babies.

Meanwhile, at least one good-quality study in the US found no advantage in starting solids early in terms of a baby's overall growth. By the age of twelve months, formula-fed babies weighed and measured the same, regardless of whether solids were introduced at three or six months, and their bodies contained a similar proportion of fat and muscle. Even though they drank slightly less milk, the overall number of calories the babies consumed was roughly the same, suggesting that babies are able to regulate the amount they eat according to how hungry they are.

Other arguments in favour of introducing early solids include exposing babies to a wide range of flavours so they're less likely to become fussy eaters, and rumours that babies will experience feeding problems if they don't learn to chew before six months of age. There's little evidence to support either of these claims, and a good-quality trial from Honduras found no difference in appetite or food acceptance between breastfed babies given solids at either four or six months of age.

On the other hand, at least one study has found that waiting to introduce solids until six months reduced the risk of asthma in babies with a family history of allergies. It's also true that babies are less likely to encounter bacteria that could make them sick if they continue to be fed milk rather than solid food.

On balance, it seems there's little benefit in introducing solids before six months, especially if you're breastfeeding. If nothing else, weaning can be hard work because it requires extra levels of organization and preparation. But if you do start after four months – particularly if your baby is showing a strong interest – there's also scant evidence that this will do any harm.

116

How do I get my baby to like vegetables?

VARIETY SEEMS TO be the key. Many babies naturally shy away from bitter flavours such as those found in certain vegetables, but if your child initially rejects broccoli or spinach, don't despair. Feeding them a range of different vegetables may be enough to get them used to the overall taste, ultimately resulting in the acceptance of the scorned vegetable.

Julie Mennella and her colleagues at the Monell Chemical Senses Center in Philadelphia took seventy-four babies aged between four and nine months with at least two weeks' experience of eating cereals or fruit and split them into two groups. The first was introduced to puréed green beans, while the remaining babies were given puréed pear. All of the babies were assessed on how well they accepted this new flavour.

Over the next eight days, the babies were further subdivided. In the bean group, some babies were given green beans to eat every dinner-time, while others were given a different vegetable at each dinner. Still others were given a variety of vegetables both within and between meals. A similar strategy was employed with the pear group, only introducing a mixture of different fruits instead of vegetables. After the eight days were up, both groups were given pears or green beans to taste once more. Although all the babies ate more of the original fruit or vegetable than they did the first time around, babies that had been eating a wide variety of different fruits or vegetables during the week showed the greatest increase in the amount of pear or green bean they consumed.

Mennella thinks that repeatedly exposing babies to lots of different flavours from the same overall food group may teach them to be more accepting of individual flavours within that group. She also suggests persevering even if babies grimace when they are eating the food – the important thing is that they are getting a taste of it. Other studies have hinted that babies often need to taste a food eight to ten times before they willingly start to accept it – although forcing babies to eat something they don't want to can result in them developing a long-term aversion to it.

If this strategy doesn't work, you could also try mixing the offending food with something the baby does like. Although this hasn't been tested in very young babies, Elizabeth Capaldi of Arizona State University in Tempe has shown that children aged two to five can be taught to like the sour taste of grapefruit juice by having it mixed with sugar the first few times they try it. Importantly, they go on liking unsweetened grapefruit juice when they taste it again, even after a gap of several weeks.

If all else fails, try showing your baby pictures of the food you want them to eat before they eat it. Several studies have suggested that toddlers are more likely to try a new fruit or vegetable if they have previously seen it in a book. A copy of *The Very Hungry Caterpillar* should definitely be at the top of any fussy eater's reading list!

117

Is there any evidence for baby-led weaning being better than parent-led weaning?

THE FIRST TASTE of food is a rite of passage for any baby (and excited parent). But quite aside from the debate about when to introduce it, the other big question is how. One current trend is to ditch spoon-fed purées in favour of baby-led weaning, encouraging babies to self-feed from the outset.

In its purest form, this means no spoon-feeding whatsoever. Babies eat what they fancy, and if that's nothing, then fine. But a baby on food strike can be a source of enormous anxiety to new parents, particularly as few of us have any clue as to how much they should be eating (see 119: 'How many calories does a baby need?'). Self-feeding can also be an extremely messy affair, and many parents worry that young babies will choke if given chunks of food rather than easy-to-swallow purées.

In my case, I had no choice: our daughter, Matilda, refused to let a spoon pass her lips unless she was controlling it (or it contained fruit yoghurt). Since she wasn't particularly adept at levering food on to a spoon (and no baby can live on Petits Filous alone), this meant that finger foods such as toast, omelette, pasta, fruit and vegetables were her main source of solids in the early days.

But is one method any better than the other? Baby-led weaning is a relatively new phenomenon, so unfortunately there's very little research to draw on, although, in 2012, British researchers published a study comparing the eating habits of 155 children aged twenty months to six and a half years who had either been introduced to solids through spoon-feeding or baby-led weaning. Key questions included what they liked to eat now, how fussy they were and how much they weighed.

In general, children introduced to solids through baby-led weaning had a stronger preference for carbohydrates than the spoon-fed children, who liked sweet foods best. The baby-led group also had a lower body mass index on average than the spoon-fed group, although there was no difference in the number of fussy eaters between the two groups.

However, one flaw of this study is that it relied on parents reporting what their children ate rather than measuring it directly. We also don't know how strict they were about the weaning process (as many parents who spoon-feed also give their children finger foods to eat), or at what age they started weaning.

Most advocates of baby-led weaning suggest waiting until the baby is six months old before introducing solid food as, before this, babies may not possess the coordination to grasp food, chew and swallow it. But there's also fairly unanimous agreement that from six months babies start needing the extra nutrients solids provide, so a key question is whether all babies are developmentally ready to start eating finger foods at this age.

In a separate study, another team of British researchers investigated the age at which babies started reaching out for food. Six hundred and two parents were interviewed when their baby was eight months old. Of these, just over half said their child had started reaching for food before the age of six months, but 6 per cent were still not reaching out for food at the age of eight months. Asked whether their babies were now eating finger foods, more than 90 per cent said they ate some at least daily, although 35 per cent were described as still needing to be fully fed at mealtimes.

American studies also hint that some babies may not be ready to eat finger foods by six months. One study found that 68 per cent of four- to six-month-olds had begun grasping foods, but just 53 per cent of seven- to eight-month-olds could eat food that needed chewing.

The message I take from all this is that it largely comes down to the individual baby. If they seem ready to start eating finger foods at six months, great, but there's currently no strong evidence to suggest that offering a mixture of finger foods and purées is any less good. It's also pretty clear that some babies won't be ready to eat finger foods at six months, so, given that they do start needing some solids from this age, it's probably wise to spoon-feed them until they are ready to self-feed.

If you do decide to try baby-led weaning, you may find that mealtimes take longer – although this isn't necessarily grounds for concern. Another study showed that fifteen-month-olds eating a finger-fed meal took around twice as long to eat half as much food as during a spoon-fed meal, but the average number of calories they consumed was the same. This is because finger foods such as toast, cheese and pasta often tend to be denser than when food is fed as a purée.

118

Should I force my baby to eat?

ONE THING BABY-LED weaning has in its favour is an altogether more relaxed attitude to food and feeding. In a study of 702 mothers, those following the baby-led approach reported feeling less pressure to get their child to eat and also seemed less concerned about their baby's weight than women who were spoon-feeding their babies. The same study found no association between weaning style and the baby's weight.

Regardless of how you choose to feed your baby, a relaxed approach is important because a domineering attitude to meal-times can have a negative impact on a baby's weight and eating habits as it grows older. Studies in children over the age of one have showed that pressuring them to eat increases pickiness, fussiness and the risk of them being underweight, while restricting access to certain foods often results in them eating more of that food when the opportunity does arise.

Although it is less well studied, parental attitude seems to affect younger babies' eating habits too. One study of six-month-olds who had been slow to put on weight found that babies whose mums had a relaxed approach to feeding put on significantly more weight in the following six months than those whose mums had a controlling attitude to food. The converse was true in babies who were heavier than average at six months. Left to their own devices, it seems that babies will eventually respond to their body's needs and regulate their food intake appropriately.

119

How many calories does a baby need?

FEW PARENTS THINK about the number of calories their baby is consuming when all they're drinking is milk, but weaning can bring on a whole new set of anxieties. I remember worrying about my daughter having eaten only a single spoon of pea purée (despite drinking plenty of milk), and trying to fill her up on banana instead – something she definitely did like the taste of.

Just how many calories a baby needs depends on its size and age, but, as a general rule, very young babies need around 110 calories per kg per day – so a one-month-old weighing four and a half kilograms would need around 495 calories (all from milk).

As babies get older their rate of growth slows, so the number of calories they need per kilogram of their body weight begins to fall. By the age of six months (when most babies start eating solid food), it has dropped to around 80 calories per kilogram per day, and it hovers around this point for the rest of the first year. This means that a six-month-old weighing seven and a half kilograms (16.5lb) would need around 600 calories per day, some of which they will get from milk and the rest from solid food. Both breast- and formula-milk contain around 70 calories per 100ml, so if a six-month-old is drinking 600ml of milk per day (420 calories), they need only an extra 180 calories from solid food. That's not very much – equivalent to approximately two bananas.

A one-year-old weighing ten kilograms (22lb) will need around 800 calories per day, and their milk consumption will probably have dropped by now. Assuming they are drinking around 450ml of cows' milk per day, which contains 300 calories (there are 67 calories per 100ml of whole cows' milk), this would mean they need an extra 500 calories from solid food. That's still only about a quarter to a fifth of what the average adult should be consuming in a day.

Of course, individual babies vary in their appetites, and there may be some days when they are hungrier or less hungry than this. The best measure of whether a baby is eating enough is whether it is maintaining a healthy weight.

120

Should I avoid feeding my baby nuts or eggs to protect it against allergies?

SOME PARENTS WORRY about introducing foods such as eggs, fish or nuts into their baby's diet for fear that this might increase their risk of developing allergies or eczema. Although there is some evidence to support this, plenty of studies have found no effect, while some even suggest that the early introduction of such foods can have a protective effect.

It's quite true that allergens from foods such as nuts and eggs can find their way into breast milk. For this reason, the American Academy of Pediatrics (AAP) used to tell mothers whose babies were at high risk of allergy to avoid feeding their child eggs until they were two years of age, and to avoid fish and nuts until they were three.

However, as more studies have been published, this advice has changed, and the current scientific consensus is that there's no reason to avoid such foods during breastfeeding or weaning. In its most recent statement, the AAP said that the current evidence does not support women avoiding certain food groups during pregnancy or breastfeeding, and there's little evidence that delaying the introduction of foods such as fish, eggs, and those that contain peanuts prevents allergy, asthma or eczema.

European experts seem to agree. The European Society for Paediatric Gastroenterology, Hepatology and Nutrition recently reviewed the evidence and concluded that, in general, solids should be introduced after seventeen weeks and no later than twenty-six weeks, adding that there is no convincing scientific evidence that avoiding fish and eggs reduces allergies, even in high-risk babies. It also cautions against introducing gluten (which is found in wheat) before four months or after seven months of age, and advises to do so gradually, as there's some evidence that this strategy reduces the risk of coeliac disease, type 1 diabetes and wheat allergy.

It's also fine to add small quantities of cows' milk to a baby's food, although it shouldn't be a baby's main source of milk before the age of one, as they can't absorb enough iron from it.

The Brown Stuff

121

Are reusable nappies really greener than disposables?

MANY ECO-CONSCIOUS PARENTS shared a collective sigh of relief when in 2005 the UK's Environment Agency announced that disposable nappies were no worse for the environment than re-usable terry-towelling ones. Disposable nappies account for around 2.4 per cent of the household waste that ends up as landfill in the UK, and few parents like the thought of bags of dirty nappies clogging up the countryside for hundreds of years (even if they are buried underground). Their degradation also produces methane, a potent greenhouse gas. However, laundering reusable nappies also takes an environmental toll, as energy is needed to fuel the washing machines that get them clean.

The Environment Agency concluded that neither type of nappy could claim environmental superiority. To put it into context, over the average two and a half years that a baby is in nappies, both are roughly comparable with driving a car between

1,300 and 2,200 miles in terms of greenhouse-gas emissions and depletion of non-renewable resources.

The US-based Union of Concerned Scientists has similarly advised parents not to waste a great deal of time or energy trying to weigh up the environmental impact of the different types of nappies, because the differences aren't particularly dramatic. In its recently published *Consumers' Guide to Effective Environmental Choices*, the union suggests opting for cloth nappies if you live somewhere with shortages in landfill space, but choosing disposable nappies if you live somewhere plagued by water shortages.

However, there are some circumstances in which cloth nappies can be greener. In 2008, the Environment Agency released an updated report concluding that the eco-credentials of reusable nappies varied widely according to how they were washed. Disposable and reusable nappies might produce similar greenhouse-gas emissions based on average washer and tumble-drier use, but washing reusable nappies in fuller loads or line-drying the nappies outdoors reduced emissions by 16 per cent, making them less damaging than disposables. If you did this and used the nappies for a second child, this would cut the environmental impact by 40 per cent (although this is still only equivalent to driving some 620 miles by car). In contrast, if parents tumble-dried all their reusable nappies, this would increase their environmental impact by 43 per cent, while washing nappies at 90°C instead of at 60°C would increase it by 31 per cent.

In other words, you can cut the environmental impact of nappies by using reusable ones – so long as you're prepared to wait until you have a full load of washing, only wash them at 50–60°C in an energy-efficient machine and leave them to dry naturally.

What about the effect of different nappy types on the health of babies' bottoms? According to a recent review of twenty-eight different studies by the Cochrane Collaboration, there is little good-quality evidence to say that either type of nappy is any better than another in terms of preventing nappy rash. Studies looking at differences in scrotal temperatures between young boys wearing disposable or reusable nappies have also produced mixed results, so any claims that disposables might affect boys' later fertility should also, for now, be regarded with scepticism.

122
What about eco-friendly disposables?

NOT ALL DISPOSABLE nappies are equal – at least if you believe the packaging. There are regular nappies, and then there are 'eco-friendly' biodegradable nappies. I know plenty of mums who have opted for the eco-friendly option in a bid to assuage their guilt about not using cloth nappies, but does it make much difference?

It largely depends on which bit of 'environmentally friendly' you care most about. In terms of greenhouse-gas emissions, there is currently little direct evidence to support the use of bio-degradable disposables over regular ones. In fact, unless you plan to compost your nappies, placing a biodegradable nappy in a dustbin is a pretty futile exercise – and could make things worse.

Bambo Nature, which claims its nappies are 75 per cent biodegradable, admits that throwing them in the regular dustbin won't cause them to biodegrade: 'Any nappy that claims to be biodegradable only biodegrades if it is composted, not if it is thrown into landfill,' says its website. This is because rubbish in landfill is usually cut off from the water, oxygen and bacteria needed for decomposition to occur.

'We think that things should biodegrade in landfill, but we've taken excruciating steps to prevent them from interacting with the natural environment because, traditionally, landfills have had major problems with the contamination of groundwater,' says Adam Gendell of US organization GreenBlue, which promotes sustainability to businesses.

Even food waste and grass cuttings struggle to decompose fully in landfill, and although some newer biodegradable nappies contain chemicals designed to speed up the breakdown of the plastics they contain, this process releases methane, which is bad for the environment.

In an ideal world, it would be better if nothing went to landfill, because many of the things we throw away could be recycled or composted. 'But if we do have to send some things to landfill – and I imagine that used diapers will be one of the last hold-outs of things that do – then it's OK for them not to biodegrade,' says Gendell.

If you want to try to compost your biodegradable nappies, Bambo recommends only home composting – not sending them to communal facilities. Even then, some of the plastic parts will not biodegrade so will need to be manually removed.

Of course, there are other environmental considerations besides greenhouse-gas emissions and biodegradability. Unbleached nappies use fewer harmful chemicals in their manufacture, and nappies containing wood from sustainably sourced forests clearly have better eco-credentials than those that don't. But the greenest option of all is using cloth nappies (washed on a moderate temperature, and left to dry naturally) and then reusing them on a second or third baby.

123

Why is baby poo yellow?

THE COLOUR OF poo is mostly influenced by two pigments, bilirubin and biliverdin, both breakdown products of red blood cells. Bilirubin starts off as a yellowish pigment, but is then further broken down and oxidized into a brown pigment called stercobilin, while biliverdin is green. Mix them together and you end up with the characteristic brown of faeces, but, just like a child's paint box, the final shade depends on how much of these pigments gets added to the mix.

The key thing that determines the final colour of poo is how long it takes to travel through the digestive system. If it travels extremely quickly, it will be green; if it takes a bit longer, the colour becomes yellow, while slower stools tend to be brown.

Breast milk is very easy to digest, so it slips through the digestive system quickly. This means very little pigment is added and the bilirubin doesn't have time to get converted into stercobilin, hence the yellow colour of breastfed babies' poo. If a baby isn't getting enough of the creamy hind-milk at the end of a feed, the milk will travel through the system even faster, often resulting in watery green poos and wind, caused by fermentation of undigested protein in the lower intestines (see 89: 'What causes colic?'). Formula-fed babies produce a rainbow of different shades from green to brown, depending on the composition of the formula (see 109: 'What is in formula milk, and why does it taste fishy?').

Apart from the tar-black meconium you might notice in your newborn's first couple of nappies, babies' poo shouldn't be black. If it is, this may be a sign of internal bleeding – as are streaks of red in the stools. Meconium isn't really poo at all; rather than consisting of digested food, it mostly contains decomposed intestinal cells that have been sloughed off during the baby's development.

As you start to introduce solid food into your baby's diet, you will probably notice stools becoming harder and browner. Certain foods can also leave unprepared parents with a fright: beetroot and raspberries can result in various shades of red, while bananas can produce what look like black worms. Because of its high-fibre content, sweetcorn barely gets digested at all and comes out looking the same as it went in.

124

Why does baby poo smell like mustard?

SIMPLE ANSWER: BECAUSE, like mustard, it contains vinegar. Breastfed babies tend to have yellow and fairly liquid stools, which smell very different to adult ones. The smell is largely influenced by the bacteria living in your bowels and breast milk contains large amounts of indigestible carbohydrates called oligosaccharides, which foster the growth of bacteria called bifidobacteria and lactobacilli. When fermented, these oligosaccharides give rise to hydrogen gas and chemicals such as propionic acid, butyric acid and acetic acid – or vinegar, as it is more commonly known.

'Oligosaccharides are also thought to be responsible for the short passage time and for the stools being soft-liquidish,' says Ekhard Ziegler, a paediatrician at the University of Iowa, who has studied gas production in breast- and formula-fed babies. Formula milk contains no oligosaccharides but larger amounts of sulphur-containing amino acids, which encourage the growth of enterobacteriaceae – bacteria commonly found in adult intestines. Parents often comment that poo from formula-fed babies is stinkier, and this is because digested formula produces more sulphurous gases, such as hydrogen sulfide and methanethiol. Babies fed soy-based formula produce particularly large amounts of hydrogen sulphide, which smells of rotten eggs.

On a slight tangent, one study recently found that mums prefer the smell of their own baby's nappies compared to those of unrelated babies. One possibility is that they're simply more accustomed to their own child's smells (and blinded by love, of course). But since gut bacteria are passed from mother to baby during birth, it is also possible that their baby's poo smells more similar to their own.

125

Is baby poo 'cleaner' than adult poo?

YOU KNOW THE scenario. Your baby's nappy explodes. Torrents of molten poo trickle out of the sides, over their clothes and leak into their sleeping bag. Then, halfway through being changed, your baby performs an alligator death roll and gets poo all over its feet and hair, not to mention you. But surely baby poo is 'cleaner' than adult poo, and therefore slightly less revolting?

Not necessarily. Babies are born with sterile guts, meaning that for the first few days of life there are few, if any, bacteria coming out of the other end. But that all changes pretty quickly, and by the end of the first month there are approximately one trillion bacteria per millilitre of poo – about the same number as in adults.

How do they get there? Babies gulp down mouthfuls of bacteria from their mother's vagina during birth and pick up still more from her skin and milk during breastfeeding. These travel down the oesophagus, through the stomach and set up camp in the intestines, where they grow and flourish. This colonization continues throughout life through the foods and drinks we consume, and many of these bacteria perform useful functions such as helping us to digest food and keeping our immune systems healthy.

There is a difference in the types of bacteria found in the poo of babies and adults, with adults harbouring a more complex range than newborns. But as solid food is introduced, the diversity of the bacteria increases until it stabilizes at around one to two years, when it reaches its adult composition – and this is as unique as a fingerprint.

The type of bacteria can also vary according to how a baby was born and whether or not they are breastfed. A study by John Penders and his colleagues at Maastricht University in the Netherlands found that babies born by C-section or who are formula-fed harbour more potentially harmful and less beneficial bacteria than breastfed babies and those born naturally. This is because in a C-section the first bacteria babies are exposed to come from the hospital environment, rather than their mothers, while formula-fed babies acquire fewer bacteria from their mother's milk and skin.

While the vast majority of gut bacteria are helpful rather than harmful, both babies and adults carry potentially disease-causing bacteria in their guts as well, so you should always be sure to wash your hands after nappy-changing.

Baby Brains

126

Is the temperament of a newborn carried through into childhood and beyond?

PRETTY SOON AFTER your baby is born, you'll realize they're far from being a blank slate and already have characteristics that distinguish them from other babies you know. Some are sleepy, others seem fascinated by new objects, while others are fussy and overreact to every situation they're placed in. Are these passing traits, or can they tell us anything about the sort of people our babies will grow up to be?

Studies of twins have suggested that certain aspects of our personalities are at least partially inherited, including how emotional, active, social and impulsive we are. But these traits don't necessarily show up in infancy. When scientists talk about personality in young babies, they tend to refer to temperament – things such as the intensity of their emotions, how sensitive they are to their environment and how easy they are to soothe. In other words, these are traits that emerge at an extremely young age and seem to be independent of experience, intelligence or learning – a bit like the basic fabric of an artist's canvas. 'Personality', on the other hand, is who you are on top of your temperament – like the painting that goes on to the canvas.

One suggestion is that many babies fall into one of three temperamental 'types'. Easy babies are generally cheerful and easy to calm, adjust quickly to new situations and routines and have normal eating and sleeping patterns. Around 40 per cent of babies are considered to be 'easy'. Difficult babies tend to be emotional, irritable, fussy and cry a lot. They often have irregular eating and sleeping patterns. Around 10 per cent of babies have 'difficult' temperaments. Slow-to-warm babies are relatively inactive and tend to withdraw from new situations and people. Although they take time to adapt to new experiences, they do accept them eventually. Around 15 per cent of babies are considered 'slow to warm'.

However, not all babies fit neatly into one of these three boxes (which is why the percentages don't add up), and there's no guarantee that an easy baby will grow into an easy toddler, and so on. Temperament can change quite a lot during the first few months, as factors such as having experienced a difficult birth or being born prematurely can have temporary effects on babies' behaviour.

There are, however, some aspects of temperament that do seem to remain relatively consistent. One of the best studied is how babies react to unexpected situations. Although most babies are fascinated by new things and being taken out into new surroundings, around 15 per cent of babies are described as 'highly reactive', which means they tend to be easily distressed by new and unexpected events – possibly because they are easily over-whelmed by stimuli. One study found that highly reactive babies are more likely to have narrow faces, which might suggest some kind of genetic predisposition.

Highly reactive babies can be hard work for parents: they fuss, cry and need more soothing than other babies – to the extent that parents may sometimes feel as if they're being deliberately pushed away. As toddlers, highly reactive types tend to be slightly more shy and reserved, and there's some evidence that they grow into more introverted adults who are at higher risk of being worry-prone and anxious. On the plus side, they are also more likely to focus on details and can be meticulous planners. But they're sensitive souls who need nurturing, and are particularly vulnerable to the effects of bad parenting (see 128: 'Can parents modify their baby's personality traits?').

Once a baby has passed the nine-month mark, some other aspects of temperament emerge that can give a hint of what parents can expect as their baby begins to blossom into a child. These include how quickly babies approach new objects – thought to be a measure of sociability and extroversion – and how fearful, angry and active they are.

Fear

Although fear might sound like a negative quality, it is thought to play an important role in the development of conscience, because children are more worried about potential punishment. Fearful babies are less likely to grow into impulsive or aggressive children, but they may also be more prone to depression and loneliness.

Novelty

Happy babies who rapidly approach new objects have a tendency to grow into more extroverted and outgoing individuals, but they may also be more impulsive, show less self-control and become easily angry or frustrated.

Anger

Babies who show early signs of anger at around ten months seem to grow into active children with a positive outlook, but they can also be impulsive and easily become angry or frustrated.

Activity

Active babies have a tendency to become active children, who are extroverted and open, but they can also be disagreeable.

127

When do babies develop a sense of themselves as individuals?

ALTHOUGH SOME ASPECTS of a baby's temperament become evident during the first year, other traits, such as pride, embarrassment, shame and guilt don't tend to emerge until babies begin to recognize themselves as 'selves'. This usually happens between eighteen months and two years of age – at around the same time they start to recognize their reflection in a mirror. You can test this on your own baby by sitting them in front of a mirror and seeing how they react to it. Babies younger than a year may reach towards the mirror or smile at it, but now try removing them and putting a small dot of lipstick or paint on the tip of their nose. After a short gap, put them back in front of the mirror and see how they react. If the baby is younger than around fifteen months, it's unlikely they'll do very much, but as they get older they will begin to realize that that blob of colour in the reflection relates to their own nose, and they'll try and wipe it off by reaching for their nose (rather than the reflection). This reaction becomes more common as babies reach eighteen months to two years.

Studies have shown that once babies start to recognize their reflections, they are more prone to showing embarrassment, as they realize that others also see them as individuals.

Between fifteen months and two years babies start to use the personal pronouns 'my' or 'mine', prompting a flurry of claims on everything from your phone – 'my phone'; to your wallet – 'my wallet'; to the floor – 'my floor'. This is the start of babies beginning to recognize themselves as individuals, but it is a fairly lengthy process. Parents will notice that babies tend to be extremely reluctant to share, and like nothing better than to snatch a toy or instrument away from another child – leading to excruciating exchanges with other parents as you try to explain away your child's apparent selfishness. Although empathy and sharing may seem like obvious things to us adults, they are only possible once you recognize the existence of two selves – one's self and another's self – each with their own identity, needs, thoughts and desires. This is also known as theory of mind. Precisely when this develops is still very much up for debate, but few researchers would argue that children have it much before the age of three, and some don't think it fully develops until children are five.

128

Can parents modify their baby's personality traits?

LET'S ASSUME YOUR baby shows signs of being 'highly reactive' (see 126: 'Is the temperament of a newborn carried through into childhood and beyond?'), but you don't want them to grow into an anxious child. You shouldn't worry too much, because parents still have a massive impact on the type of person their baby grows up to be.

Let's take a Dutch study which found that babies deemed to be highly reactive or irritable when they were fifteen days old were less likely to be securely attached at the age of one. Secure attachment is a reflection of how 'bonded' a baby is to its mother or main caregiver, and babies who are insecurely attached show little preference for that caregiver over a complete stranger.

The researchers found that mums of irritable babies interacted with them less and were less likely to notice when their babies were giving them positive signals about wanting to play or be cuddled than the mothers of non-irritable infants. However, if the same researchers gave the mums advice on how to comfort and play with their babies, they found that these babies were just as likely to have formed secure attachments by the time they were one as non-irritable babies. This suggests that how parents respond to their baby's temperamental traits can play a huge role in how they develop. An irritable baby may be harder to soothe and push its mother away, but if a mother pays attention to its needs and provides comfort when it is unhappy, it will flourish just like any other child.

The culture in which a child is raised may also play a big role in how different temperaments play out in later life. One study that compared how Canadian and Chinese mothers responded to shyness in their children found that, whereas Chinese mothers approved of shyness, Canadian mums saw it as a negative trait and tried to draw the children out of their shells. A later study by the same group also found cultural differences in the way peers reacted to a shy child. Canadian four-year-olds tended to reject shy classmates, while Chinese children seemed to like them. Perhaps it's no wonder that Chinese people are often more restrained than Westerners.

129

Does the order of birth influence children's personalities?

ARE FIRST-BORNS SUCH as Tony Blair and Bill Clinton ambitious yet conventional because they're first-borns? Perhaps Ricky Gervais and Bill Gates are helpful and sociable because they're middle children, and Charles Darwin and Mozart were creative and rebellious because, as younger siblings, they felt the need to stand out.

The extent to which order of birth affects personality is a debate that has been raging between psychologists for several decades. A number of studies have suggested that first-borns tend to be more achievement-oriented, conscientious, hard-working, organized and reliable than later-born children, while later-born children are generally more extroverted, open to new experiences and agreeable than first-borns. The latest research suggests that when it comes to certain attributes, there may be some truth to it, even if much that has been written on birth order is flawed.

Although most parents believe they raise siblings in exactly the same way, the research contradicts this. Studies of identical twins have found that certain attributes of their personalities, such as intelligence, are up to 70 per cent genetic. However, if twins are split up and raised in separate families, studies have found that their social attitudes and interests, as well as personality traits such as how aggressive or impulsive they are, are no more different than if they had been raised in the same family. In other words, growing up in the same family doesn't have a great deal of impact in terms of making siblings resemble each other. This suggests that parents respond to siblings as individuals – even if they are twins.

One reason why much of the research on birth order has been criticized is that it hasn't necessarily taken the size of people's families into account. If most of the first-borns who are studied come from small families or were only children, then they're likely to have had very different upbringings to second- or third-born children from large families, regardless of when in the order of siblings they were born.

However, several recent studies have been done that don't suffer from this problem, and they provide at least some support for the idea that birth order makes a difference – albeit a small one.

For example, in 2009, Joshua Hartshorne and his colleagues at Harvard University found that people tend to associate with those who are similarly in line in the family pecking-order, so first-borns are more likely to befriend other first-borns, and so on. Since other studies have shown that we tend to be attracted to people with similar personality traits, this is at least indirect evidence that birth order might shape our personalities to some degree.

Birth order may also have an impact on intelligence. Petter Kristensen and his colleagues at Oslo University studied the military records of 241,310 Norwegian conscripts and found that, on average, eldest children had higher IQs than second-born children. It wasn't a big difference – just 2.3 points on the IQ scale – but it might be enough to put someone at the top of their class rather than the middle (and this might have other knock-on effects in terms of their confidence or how hard they're pushed to work). Interestingly, it didn't seem to come down to actually having been born first but to being raised as the elder child. Kristensen found that those who took on the role of the eldest child after an elder brother or sister died showed a similar intellectual advantage.

130
Why do twins develop different personalities?

IF ANYTHING, THE impact genetics has on personality seems to be less strong at birth than later on in babyhood. A study of identical and non-identical twins born in Kentucky found no similarity in terms of their irritability, resistance to soothing, activity level or how responsive they were to sounds and objects during their first week of life – despite identical twins sharing the same DNA. So what else could account for the differences seen between twins? Although twins share the same uterus, they don't necessarily have the same experience while they're in there. Twins are often different sizes at birth, because the nutrients they receive aren't equally distributed. One twin may have a more difficult birth than the other one and, if there are complications, then they may receive different treatment in hospital. Even once they get home, they may be fed or attended to in different orders, or by different parents. And if one twin starts to behave in a certain way – waking more frequently at night, for example – then its parents may start to treat it differently, giving it more cuddles or having less patience when it cries.

Twins do seem to become more like each other towards the end of the first year, but they're still far from identical. And as children age and have increasingly different experiences, their personalities diverge even further. But it's not just personality. You might assume that twins with identical DNA would be at the same risk of diseases such as diabetes, schizophrenia or heart disease, but they're not. Neither do they tend to die from the same cause. Just why this should be is currently the source of intense research, but a likely explanation is that different experiences cause their genes to be more or less turned on or off through a process called epigenetics.

131

Do babies like some people better than others?

SOME PEOPLE LOVE babies; others can't stand them. The funny thing is that babies may feel the same way about us. Given a choice between two faces – one attractive and one unattractive – a newborn baby will choose to fixate on the beautiful face. Alan Slater and his colleagues at the University of Exeter, UK, asked adults to rate a variety of female faces on attractiveness, then searched for pairs of photographs that were well matched for brightness and contrast but at opposite ends of the attractiveness spectrum.

These photos were then shown to babies ranging from one to seven days old, while a researcher stood out of sight and noted in which direction their eyes moved. Nearly all of the babies fixated on the more attractive face.

Slater suggests that faces we judge to be more attractive come closer to the prototype for a human face, which babies are hard-wired to recognize (see 78: 'What do newborns know?'). Previous research has shown that if you meld together lots of different faces to create a single one, the face you come up with is rated as extremely attractive by many.

But babies don't just judge books by their covers. Newborns also get upset if people's facial expressions suggest they're not responding to them. Experiments with babies as young as three hours old have shown that if their mother pulls a blank, expressionless face, the baby responds by decreasing their eye contact and showing signs of distress. This suggests that babies are born with certain expectations regarding the rules of interpersonal communication. It might also explain why babies often start to cry if someone glares at them.

By the age of six months, babies seem to have become pretty good judges of character. Karen Wynn and her colleagues at Yale University in New Haven, US, showed twelve babies a puppet show in which different toys with googly eyes climbed up a hill, either helping or hindering each other. For example, in one scenario a yellow triangle helped a red circle to climb the hill and then a blue square aggressively pushed it back down again.

After the show had finished, the babies were offered all the toys to see which one they preferred. All twelve six-month-olds picked the helpful character. They got a similar result when the same experiment was done with ten-month-olds.

When two of the blocks moved up the hill together, the babies seemed unsure of which one they preferred – possibly because they were unsure of the blocks' intentions.

This experiment suggests that babies may start to make judgements about social situations about a year earlier than previously assumed. By around eighteen months of age, toddlers will actively start trying to be helpful, picking up a broom and trying to sweep the kitchen floor, for example.

132

How much do babies remember?

WHAT IS YOUR earliest memory? Probably it was some time between the age of two and four, but you may be confused about the details and wonder if it ever really happened at all. That's how I feel about my first memory: being pushed up a hill in Canada, eating from a packet of Fruit Loops. According to my mother, we did visit Canada when I was about two and a half, and I probably did eat Fruit Loops – but whether this is an actual memory is hard to tell.

For a long time, scientists believed that babies had no memory at all – something they called 'infantile amnesia'. But that view has changed in recent years, as we learn more about the different types of memory that exist and when these begin to form.

It's quite true that if you ask an adult to recall their earliest memory, most can't remember anything before the age of two or three. There's also some evidence that emotionally charged events are the most memorable. One study that asked college students whether they could recall life-changing events that happened before they were five years old found that memories of being hospitalized or the birth of a sibling seemed to be particularly memorable, as details of these could be recalled from around the age of two, while the death of a family member or moving to a new home tended to be recalled if it happened after they were three.

This period of a child's development is marked by the emergence of two other skills that may help to explain why these 'autobiographical memories' don't seem to stick from an earlier age. One is a sense of 'self', as autobiographical memories are characterized by the fact that they usually place you as an individual at the centre of them. The other is language. Several studies have suggested that possessing the language to describe events and objects may help us to remember them. For one thing, it may provide children with extra prompts for accessing memories. Language also provides opportunities for rehearsing and reviving memories through conversations with family and friends. It may also help in the laying down of memories in the first place – as many parents notice how two- or three-year-olds sometimes lie in bed and seem to chatter through the day's events to themselves.

The importance of language in summoning up memories was shown through an experiment using something called 'the magic shrinking machine'. Two- to three-year-old children were asked to take a toy from a box and put it into the top of the machine. They then pulled a lever and a miniature version of the same toy appeared from a chute at the bottom of the machine (there was an adult sitting behind the machine, who was selecting the appropriate toy to release from the chute). The children's language ability was assessed when they first saw the machine, and again when they were asked about it either six months or a year later. When the children were asked what they remembered about the machine and how it worked, they were incapable of describing things they didn't have the appropriate vocabulary for at the time of first seeing it – even if their language ability was now very good. They simply couldn't find the words to describe it, even if they possessed those words now. However, it was clear the children remembered the machine, as they could use photographs to explain how it worked.

133

Why don't we remember being babies?

JUST BECAUSE ADULTS can't remember their early childhoods, this doesn't mean that babies don't remember what happened to them from one day to the next. It's clear that even foetuses have the ability to learn, which requires some degree of memory. For example, newborns seem to recognize pieces of music that were played to them and individual voices they heard before birth (see 36: 'What do babies learn during their time in the womb?'). This is different to autobiographical memory, but it is memory nonetheless.

In general, the older a baby is, the longer it will hold on to memories. Some of the most famous experiments investigating memory in tiny babies were done by Carolyn Rovee-Collier, now at Rutgers University in New Jersey. She took babies as young as two months and attached one of their feet by a ribbon to the brightly coloured mobile that hung above their cot. If the babies kicked their feet the mobile would move, and if they kicked really hard the figures would knock against each other, producing a clanking sound – a very exciting thing for a nine-week-old.

After playing this game for two nine-minute sessions, the babies were tested days later to see how long they would remember that they had to kick their feet in order to get the mobile to move. Two-month-old babies held on to this knowledge for just one to two days, but six-month-old babies could remember it for two weeks.

How many times they got to play the game made a big difference to how long they remembered it, however. If the two-month-olds were given three six-minute training sessions, for example, they could remember the game for two weeks.

One theory is that babies have no problem laying down memories – the problem is retrieving them again afterwards. This could also explain why we adults struggle to access memories from our early years: our view of the world has changed so much that the simple cues that once triggered memories – that enormous, looming chair, or the colourful buttons on our mother's jumper – are no longer present or no longer carry the same salience as they did when we were tiny babies.

In a later experiment, Rovee-Collier found that periodically reminding babies of the mobile game by simply showing them the mobile (but not actually attaching their foot to it) extended their memory of it. Three-month-olds given two nine-minute training sessions generally couldn't remember to kick their feet when they were tested two weeks later, but if they were shown the same mobile twenty-four hours before the test, they could. It's as if seeing the mobile enabled them to replay the game in their heads, so that when they were tested twenty-four hours later, they knew exactly what to do.

But these kinds of experiments test only a baby's ability to remember skills or habits, which isn't the same as consciously remembering experiences such as a trip to the circus or the cake your grandmother baked for your first birthday. Obviously, this type of memory is difficult to assess in babies, because they can't tell us what they can remember, but Andrew Meltzoff of the University of Washington in Seattle has found an indirect way of testing it.

He decided to show babies unexpected ways of using several new objects, such as a cup that collapses if you press down on it from above and a box that lights up when you touch it with your forehead. The question was whether babies would remember what to do with these objects if they'd watched an adult using them but never handled the objects themselves.

When he did this with fourteen-month-old babies, he found that when they were given the objects two to four months later, many of them immediately picked them up and imitated what they had seen the adults do with them. Similar studies have hinted that babies as young as nine months may also have this ability to remember things without directly experiencing them first.

None of these memory tests are perfect, and they don't necessarily reflect real life, when babies may be exposed to objects or experiences repeatedly, or in different situations and contexts, but they certainly suggest that babies have strong memories – even if they can't recall them when they grow up. It is also un-clear how modern technologies such as digital cameras and video might influence children's ability to remember things. When my daughter, Matilda, was around twenty months old, she became obsessed with watching a recording of some circus horses and their trainer, Yasmine Smart, whom we'd seen performing at Zippos Circus in London several weeks earlier. After visiting the circus, she asked to watch the video almost every day. I'm fascinated to know whether she will retain the memory of that circus visit for many years to come – although I will have to ask her about other details of the circus, besides the horses, to be sure she isn't just remembering the video.

134

Does nursery make babies sociable or stressed?

SHOULD YOU DECIDE to go back to work, choosing childcare for your baby is likely to be one of the most stressful decisions you're going to have to make. Newspaper headlines warning that babies who attend nursery experience damaging levels of stress hormones only serve to make parental guilt worse, yet many of us have little option if both parents wish to pursue their individual careers. On the other hand, there's also the nagging feeling that playing with other children is likely to be good for a baby's social development, so maybe nursery is a positive thing after all.

Cortisol is a hormone that is released during normal daily activity, but it is also released in large amounts in response to stress. Several studies have found that children attending full-day childcare have higher levels of cortisol on the afternoons that they attend childcare compared to days when they are at home with their parents (a pattern also seen in adult executives working in high-stress jobs). For example, one study of toddlers aged eighteen months to three years old found that afternoon cortisol levels were one and a half times higher on the days the toddlers attended nursery, even though the centres they attended scored highly on measures of childcare quality.

But is stress an inevitable consequence of being separated from one's parents? To investigate, Andrea Dettling of the Swiss Federal Institute of Technology compared three- to four-year-olds who were placed in nurseries for around forty hours a week with those who spent similar amounts of time away from home with a childminder, and with those who received no external childcare. It was a small study, with just twenty children in each group, but it did find that being separated from one's parents for the day isn't necessarily stressful in itself. Although children attending nursery showed greater release of cortisol in the afternoons (as had been seen in previous studies), cortisol release among children being looked after by childminders depended on the amount of individual attention the child received from their carer. Those whose childminders gave them plenty of individual attention showed no difference in cortisol levels compared to when they stayed at home, while those who received lower-quality childcare showed patterns similar to those children attending nursery.

Even if nursery does raise children's cortisol levels, this doesn't necessarily cause any long-term damage. One of the largest studies to address these issues is the Study of Early Child Care (SECC), which is funded by the US National Institute of Child Health and Human Development. It has followed the development of more than 1,200 children from birth through to middle-childhood.

On the whole, it has found that children who are cared for exclusively by their mothers do not develop differently to those who are also cared for by others. Those who receive high-quality childcare also get some additional benefits, including slightly better cognitive function and language development across the first three years of life and greater school readiness at four and a half years of age – although the effect is small compared to the influence that individual parents have on their children's development. This means that children whose parents don't have the time or inclination to interact with them may benefit from attending nursery, but the benefit is likely to be small if they receive such stimulation at home. In terms of boosting young children's cognitive and language development, the most important thing of all seems to be asking children questions, responding to their attempts to speak and engaging in other forms of communication, such as singing songs, reading books and getting them to repeat numbers, shapes or letters.

It's a similar picture when it comes to children's social and emotional development. The SECC study has found that quality of parenting has a far greater impact on this aspect of development than whether children are regularly cared for by someone else, or on the quality, quantity or stability of this care. That's not to say that childcare has no effect: higher quality care results in greater social competence, cooperation and less problem behaviour at two and three years of age, compared to lower quality care.

It's also true that greater experience in groups with other children (regardless of whether they get this through attending nursery or by going to toddler groups with their parents) makes young children slightly more cooperative and better at interacting with others as they get older. Other studies have hinted that children who attend childcare are more accomplished at entertaining themselves, have greater self-confidence and show less distress in new situations. The amount of time children spend in care also has an effect. Being in childcare for thirty or more hours a week is associated with a small but statistically significant increase in behavioural problems.

Finally, it's worth noting that some children appear to be more affected by poor-quality childcare and get more out of high-quality childcare than others. In particular, babies who are fussy, difficult to soothe and irritable tend to experience more behavioural problems when faced with low-quality childcare but fewer behavioural problems than those with easy temperaments if they experience high-quality childcare. These difficult babies may even be rated as more socially competent than children with easy temperaments, provided they receive sensitive and encouraging care from their minders.

So how can parents judge the quality of care that a nanny, childminder or nursery is offering? Generally speaking, higher levels of education and training among caregivers seem to be associated with better quality care, as does having a low ratio of children to adults and a small overall group or class size. For babies aged six to eighteen months, the American Academy of Pediatrics recommends having no more than three babies to one adult, and limiting overall group size to six babies. For eighteen- to twenty-four-month-olds, it suggests no more than four toddlers to one adult, and limiting groups to eight.

Researchers in the SECC study have also concluded that high-quality childcare hinges upon sensitive, encouraging and frequent interactions between the caregiver and the child – something that may be harder to achieve within larger groups, or if a caregiver has too many children to look after.

In order to assess this, they recommend that parents look out for the following signs when choosing childcare, all of which are markers of high-quality care:

- Is the caregiver generally in good spirits and encouraging when interacting with the child?
- Does the caregiver hug the child, pat the child on the back or hold the child's hand? Does the caregiver comfort the child if they are upset?

- Does the caregiver repeat the child's words, comment on what the child says or tries to say and answer the child's questions?
- Does the caregiver encourage the child to talk/communicate by asking questions that the child can answer easily, such as 'yes' or 'no' questions, or asking about a family member or a toy?
- Does the caregiver respond to the child's positive actions with positive words, such as 'You did it!' or 'Well done!'? Do they encourage the child to say the alphabet out loud, count to ten and name shapes or objects?
- Does the caregiver tell stories, describe objects or events or sing songs?
- Does the caregiver encourage development by, for example, giving young babies tummy time to help neck and shoulder muscles grow stronger, helping older babies to walk or to finish puzzles, stack blocks or zip zippers?
- Does the caregiver encourage the child to smile, laugh and play with other children? Do they support sharing between the child and other children and provide examples of good behaviour?
- Does the caregiver read books and stories to the child? Do they let the child touch the book and turn the pages?
- Does the caregiver ensure they are positive, not negative, in their interactions with the child – even when the child is misbehaving? Does the caregiver make it a point to interact with the child and not ignore him or her?

135

Are babies born in summer any different to winter babies?

ALTHOUGH YOUR BABY may not show the classic indecisiveness of a Libran or the adventurous spirit of a Sagittarian, consider this: if they were born in the UK between September and December, they are twice as likely to become a professional soccer player as a baby born during the summer months. Unfortunately, they're also slightly more likely to suffer panic attacks and, if they're male, to become an alcoholic when they get older. Meanwhile, cricket fans may want to plan on giving birth early in the year, as fast bowlers are more likely to have been born in late winter and early spring, while if you're ambitious for your child to become a medic, you might aim for an April–June birthday (at least if you live in the northern hemisphere).

There is also some evidence that a baby's season of birth can influence its health. For example, babies born in the northern hemisphere between December and April are more likely to develop schizophrenia or bipolar disorder than those born at other times of the year, while babies born between April and June are at greater risk of developing anorexia.

The effect of your birth month on all of these things is small, of course, but it is too great to be simply dismissed as chance. The risk of schizophrenia (which is among the best studied of these season-linked disorders) for babies born in the northern hemisphere during February, March and April is around 5 to 10 per cent higher than for babies born at other times of the year. This is similar to the increased risk associated with having a parent or sibling with the disorder.

Scientists have also proposed some mechanisms by which these trends might come about. In the case of sport, something called the 'relative age effect' has been used to explain why babies born at certain times of the year are more likely to become professional players. For example, selection for national youth teams takes place between September and December in the UK and, according to the Football Association, 57 per cent of players at Premier League academies in the 2008–9 season were born between these months, compared to just 14 per cent of children born between May and August. The proposed explanation is that a September baby will have had a year longer to practise and may be physically stronger than a child born in August. In other parts of the world, including much of Europe and North America, selection for soccer teams takes place during the early months of the year, putting spring-born babies at a greater advantage.

Different factors are likely to be at work when it comes to health. Some studies have suggested that the increased risk of schizophrenia among babies born in the springtime is the result of their mother being exposed to viruses such as influenza during a critical period of pregnancy. More recently, scientists have proposed that vitamin D deficiency caused by low levels of sunlight during pregnancy may be to blame, or that varying levels of brain chemicals such as serotonin or dopamine at different times of the year might play a role.

In some countries, a baby's season of birth can also influence their physical development. The city of Denver, Colorado, nestles in the heart of the Rocky Mountains, so it experiences some pretty harsh winters. A study of 425 babies living there found that those born in summer and autumn started crawling around three weeks later on average than babies born in winter. Babies typically start to crawl at between six and eight months of age – the height of winter for an American baby born during the summer months. Fewer daylight hours, not to mention the snow and biting cold, mean that such babies will probably spend more time bundled up in a cosy buggy rather than getting out and practising their fledgling crawling skills. By contrast, babies born in winter are learning to crawl at precisely the time when they are most likely to have the space and motivation to explore their surroundings: summer. Other studies have confirmed that physically active babies who start to practise crawling or stepping actions earlier also tend to be the first ones to get on the move – usually much to the consternation of their frazzled mothers.

136

Does being born at the start of the school year give you an academic advantage?

IT MIGHT DO, although the effect diminishes with age. Numerous studies have found that children born at the start of the academic year tend to score higher in tests than children who are the youngest in their year group. One of the largest studies to look at this issue was conducted in England, where the academic year starts in September, making August-born children the youngest.

By the age of seven, August-born children are two to three times as likely to be rated as 'below average' by their teachers in reading, writing and maths. For example, 14 per cent of September-born children were rated 'below average' on their reading ability, compared to 30 per cent of August-born children. The gap does seem to narrow as children age, though, suggesting that August-borns do make up some of the lost ground.

Overall, around half of children in England attain five GCSEs (exams taken aged sixteen) at grades A to C, including English and maths. However, if you compare children born in August with those born the previous September, girls are 5.5 per cent less likely to hit this target, while boys are 6.1 per cent less likely to do so. The same researchers found that August-born teenagers are 20 per cent less likely to get into top universities such as Oxford or Cambridge than their September-born counterparts. Despite all this, the researchers found no difference in children's sense of self-worth, their enjoyment of school or their expectations of and aspirations for further and higher education.

The jury is still out on what's to blame for the disparity – or what to do about it. One possibility is that August-born children simply aren't ready to start school at the age of four and so don't benefit from formal education during this time; another is that it's down to making children sit the tests on the same day, regardless of their age.

Another problem is that such studies don't take the abilities of individual children into account. For example, we don't know if August-born children with high IQs would be similarly affected by their date of birth or if this would buffer them from the effect. Neither is it clear whether children who attend nursery or whose parents spend a lot of time reading to them before they start school are similarly disadvantaged.

What is clear is that there are plenty of August-born geniuses to testify that there are always exceptions to the rule.

Language

137

When will my baby understand what I'm saying?

MOST CHILDREN BEGIN to utter their first words around their first birthday. Until very recently, babies younger than ten to twelve months were referred to as 'pre-linguistic': although they could recognize the common vowel and consonant sounds produced by those around them, they couldn't understand the meaning of actual words. A popular theory was that babies couldn't grasp language until they started viewing other people as individuals with their own motivations – which generally begins around nine to ten months. Before this, they didn't understand enough about people's intentions to associate these sounds with actual things or actions.

However, new research is starting to challenge that view. Elika Bergelson and Daniel Swingley at the University of Pennsylvania in Philadelphia tested the ability of six- and nine-month-olds to link words for food and body parts to the appropriate pictures by prompting them to 'look at the apple' and then tracking their eye movements. They found that babies as young as six months were capable of identifying the correct object – including when it formed part of a picture containing several objects, such as a kitchen table with an apple, biscuit, yoghurt and bottle on it.

The message we can take from this is that even young babies may understand some of what is said to them, so it makes sense to talk to them as much as possible. Research on older babies has also shown that engaging babies in conversations is a far more effective way of teaching language than leaving them to absorb it passively, even if they don't answer back yet.

138

Why do babies say 'Dada' before 'Mama'?

LEARNING TO SPEAK requires the coordination of more than seventy different muscles and body parts and is one of the most complicated physical actions we humans perform. So perhaps it's no wonder that fully mastering language takes children around five years.

At around seven or eight months of age, babies start to babble. Barbara L. Davis at the University of Texas in Austin travels the world trying to capture some of these first utterances and make sense of them. Using tiny microphones attached to babies' bibs, she has recorded babble from as far afield as China, Ecuador and Romania. She's also heard American, French, Dutch and Berber babies babble. 'What we've found is that children are extremely alike during this babbling period,' she says.

Sounds such as 'b', 'd', 'm', and 'g' combined with 'ah' seem to be the easiest to make, so that's where babies start. 'You open and close your lips and you get "ba ba ba",' says Davis. Babies can't control their tongues very well, so cannot articulate other sounds, but they can move them forwards and backwards. 'If you put your tongue at the front of the mouth you get "da da da", and if you move it to the back you get "ga ga ga",' says Davis.

At this stage, babies are not necessarily trying to communicate, they are simply producing rhythmic sounds which will provide the foundation of later language. Although many parents assume that 'Mama' and 'Dada' are among a baby's first words because these are the people who are most important to them, baby babble isn't thought to start taking on meaning until around ten months to a year. So if your baby says 'Dada' before 'Mama', it doesn't necessarily mean he likes Daddy best.

Many babies do tend to say 'da' before 'ma', though. 'Early speech sounds tend to be produced by moving the lips, jaw and soft palate simultaneously,' says Jordan Green at the University of Nebraska in Lincoln. This is how we make 'ba' and 'da' sounds, whereas 'ma' requires us to move the lips and jaw but relax the soft palate so the sound enters the nasal cavity.

Adults also use 'da' and 'ba' more frequently than they use 'ma' and 'na', and babies seem to have figured this out. '"Da" is the most common sound across the world's languages,' says Davis. 'Probably the reason they are saying "Dada" sooner than "Mama" is because they are making more of these oral sounds.'

It could even be that the very reason these words exist is because, throughout history, parents have seized babies' first utterances and used them to label themselves. Indeed, the baby words for 'mother' and 'father' are remarkably similar across languages. Father is 'dada', 'papa', 'baba' and 'abba' in English, French, Chinese/Persian and Aramaic respectively; while 'mama' is the same across English, Chinese and Swahili, and is 'haha' in

Japanese. There are exceptions, such as 'chichi', which is Japanese for father, but these seem to be in the minority.

Some of the hardest sounds to make are 'r', 'l', 'ph', 'ch' and 'sh', so these tend to be some of the last sounds that children master. This explains why many toddlers say adorable-sounding words like 'babbit' instead of 'rabbit'.

139
When does baby babble take on meaning?

ALTHOUGH BABIES START babbling at around seven months, most language experts believe it takes at least a few more months before baby babble morphs into something more meaningful. On average, babies say their first words at around the age of one, although there is enormous variability in when babies start to speak. Some babies don't babble at all and move straight into speech. 'There are children who are relatively more or less voluble, but it's certainly something to pay attention to if there's none of it, or if they continue making "buuuuuuuh" or "muuuuuuuh" sounds that aren't very rhythmic and syllable-like,' says Barbara Davis.

Recent studies have found that babbling may be more than simple mouth movements, however. You may never have noticed it, but when people talk, the right side of their mouth opens a tiny bit wider than the left. This is because the left side of the brain (which controls the right side of the face) plays a greater role in producing language. To investigate whether babbling might be a first shot at language, Laura-Ann Petitto and her colleagues filmed ten babies aged between five and twelve months while they were babbling, making other noises or smiling. She then slowed down the tapes and asked people who knew nothing about the study to rate which side of the babies' mouths opened more.

Sure enough, when the babies were babbling they showed the same pattern of mouth-opening as adults, suggesting that the language areas of their brains were being used. Smiling had the opposite effect, which may be because the right side of the brain is more involved in processing emotions.

When they do appear, first words often contain many of the same sounds as early babbling, which makes it even more difficult to detect precisely when babbling ends and words begin. Parents are generally better than strangers at spotting the transition, because they spend every day with their babies and can spot patterns such as saying 'ba' every time baby gets into the bath.

Although words for people and objects are relatively easy to identify, studies suggest that several non-object words such as 'gone', 'there', 'uh-oh' and 'more' are also among a baby's early vocal repertoire. During the 1980s, Alison Gopnik of the University of California in Berkeley recorded one-year-olds going about their daily business in order to find out what words they were using and how they were using them. In her book *How Babies Think* she describes how an eighteen-month-old called Henry repeatedly used the word 'gone' to describe the many and varied ways in which objects disappear from view, rather than meaning that something is finished, like dinner.

Similarly, babies often use 'oh dear' or 'uh-oh' to describe failures, and 'there' to indicate success. They also extend naming words to other people or objects with similar characteristics, so male friends and the postman also become 'Dada', and the definition of 'moon' is extended to include all manner of lamps, oranges and fingernail clippings.

My daughter has taken to using 'bumpy head' whenever she is upset or frustrated. Clearly, she associates crying with a time when she bumped her head and feels that this is the most appropriate way to describe her emotions.

140

Do babies communicate without language?

BABIES CLEARLY UNDERSTAND that crying elicits attention, but they soon develop other means of communication as well. Stick your tongue out at a one-month-old and often they'll stick theirs out back. Talk to a young baby and you may notice them stop moving and then responding with a volley of wriggles, coos and smiles. All this suggests that even young babies understand the fundamentals of communication and they get better at it with every passing day.

Babies also love having their own actions mimicked. One study found that three-and-a-half-month-old babies cooed and smiled more frequently when mums imitated their sounds and expressions compared to when they responded with different sounds and actions.

As babies progress from cooing to babbling, they begin to realize that they can use these sounds to get adults to interact with them – a bit like starting a conversation. In a recent study, Michael Goldstein and his colleagues at Cornell University in Ithaca, New York, studied how five-month-olds reacted when an adult they had been playing with suddenly assumed an expressionless face and started ignoring them. Babies hate this. If you do it to a younger infant, they will invariably look away and start to cry.

But these five-month-olds reacted differently. Rather than crying, they increased their coos and squeals, apparently trying to engage the adult in conversation. When they got nothing back, they eventually stopped trying and just looked sad.

At around the age of one, babies also start pointing. The traditional view is that they do this only to get things they want, but more recent studies suggest that one-year-olds also point in order to be helpful. This is surprising, because helpfulness implies an understanding that other people have minds of their own and don't necessarily know everything you do – a complex social skill that wasn't thought to develop until later.

Ulf Liszkowski and colleagues at the Max Planck Institute for Psycholinguistics in Nijmegen, the Netherlands, sat one-year-olds at a table and did craft projects with them. For example, they took an ink stamp and used it to print on a piece of paper, then took a roll of sticky tape and used it to stick the edges of a piece of cloth together. Then they did it again, but this time they deliberately knocked the roll of tape off the edge of the table, down a chute and on to the floor. When it came to the moment when the tape was needed, the adult began to hunt for it. In nearly every case, the one-year-olds pointed to the dropped tape in order to help them find it.

This suggests that one-year-olds understand others' ignorance and want to help them. Even though they can't yet say 'You've dropped something', they can tell them by using this newly acquired pointing skill.

141

Do more physical babies develop language faster than sedentary babies?

ANY PARENT WILL tell you that babies are different. Some are more sociable, others more active, but it seems the very act of starting to move about marks a change in the way they interact with other people. Several studies have shown that once babies start to crawl, they show both more anger and affection towards their parents, more sensitivity to their comings and goings and pay more attention to what's going on around them. What's more, it seems to be the act of movement that triggers these changes, rather than the other way around. If you put a baby who can't yet crawl into a baby-walker, it behaves very differently to when it is left lying on a play-mat. When they can't move, babies are less sociable and mostly focus on items on the floor. But put them in a baby-walker and they behave just like mobile infants of the same age, engaging with adults in the room and looking at things on the walls. It seems that this new perspective on the world somehow stimulates them to become more sociable. Possibly, it's simply that they can see more once they've started crawling or are upright.

The transition to walking also seems to trigger a change in babies' social behaviour. Melissa Clearfield of Whitman College in Walla Walla, Washington, compared nine- to twelve-month-olds who could walk with babies of the same age who were still crawling but could stand upright and move about the room in a baby-walker. She found that babies who could walk independently spent more time interacting with toys and their mothers. They also made more attempts to babble or speak, and more gestures, such as pointing or waving at a toy while looking at their mothers. Importantly, the babies in the baby-walker had their hands free and so could have done the same things, but they didn't. It seems the transition to independent walking itself changes how babies interact with others.

142

Why do women speak to babies in a silly voice?

'Coochy-coo.' 'Want Mama to play with that?' Let's face it, we've all caught ourselves speaking like idiots when a baby is around, but there's a genuine reason for doing it: babies like it. Given a choice between listening to a mum talking to a baby and the same mum talking to another adult, babies will choose the baby talk. It even has a proper name: 'motherese'.

Parents across the world show a common pattern when talking to babies. They raise the pitch of their voice, speak more slowly, elongate the vowels and exaggerate changes in intonation, giving their speech an almost musical quality. Sentences are shortened and subtle variations of the same sentence are often repeated.

What's more, it seems that this unconscious shift to motherese may help babies to learn language. Previous studies have shown that English, Swedish, Chinese and Russian mothers unconsciously tailor their motherese to make the vowels of their language particularly resonant, with phrases such as 'Oooooh, you're so cuuute, with those biiig bluuue eeeyes.'

More recently, Jae Young Song and his colleagues at Brown University on Rhode Island tested nineteen-month-olds and found that this elongation of vowel sounds significantly enhanced their ability to recognize words.

It's not only our voices. Recent research has shown that we change our behaviour too. Mothers were recorded showing a series of toys to adults and babies, including a snake made up of a string of coloured wooden beads and a clear plastic hamster ball that contained a smaller coloured ball inside it. Compared to when they showed adults how these toys worked, the women were more interactive, enthusiastic and repetitive when they demonstrated them to babies. What's more, given a choice between watching the adult-directed demonstration and the baby-directed one, babies were far more interested in the one that mums had instinctively tailored to babies.

Dads also seem to change their physical behaviour when they interact with babies, but in a slightly different way. When dads demonstrated the same toys to infants they were slightly less enthusiastic than the mothers, but they seemed to get closer to the babies and allow them more opportunities to touch and explore how the toy worked. This desire to demonstrate the intricacies of how things work has also been seen in studies that have looked at how fatherly behaviour is influenced by hormones such as prolactin (see 32: 'Do men change when they become dads?'), so I wonder if it's a more general difference between men and women.

Such changes in physical behaviour might be important in capturing babies' attention, as well as teaching them about human motivation and how to explore, test and understand the things they come across in everyday life.

143

When do babies start learning language?

BABIES ARE BORN with the ability to absorb any language, but even before birth they are beginning to sift through the various sounds and rhythms they hear and trying to make sense of them. Four-day-old babies can distinguish the language they've heard in the womb from other languages (see 78: 'What do newborns know?'), although it may take up to a year for them to utter their first word.

Even though babies start categorizing sounds and rhythms from birth, for around nine months their brains remain 'international' in the sense that they can discriminate all 150 speech sounds that occur across human languages. Think of adult Japanese speakers, who struggle to differentiate between the 'r' and 'l' sounds of English, or English speakers who often fail to perceive the difference between 'p' and 'b' in Spanish and French. None of this is a problem for a young baby, who can also detect these differences regardless of who is talking to them.

When Patricia Kuhl of the University of Washington in Seattle set out to test this universal language ability of babies across continents, she found a curious pattern. At some point during the second half of the first year, it switches off. Japanese babies that were perfectly capable of perceiving the difference between 'r' and 'l' at seven months (as measured by their tendency to turn their head towards the loudspeaker when the sounds changed) were incapable of discriminating between them by eleven months of age. It's as if the brain gradually starts to filter out sounds that are irrelevant in everyday life.

Whether it's possible to regain this universal sound perception is a matter of debate. For a long time the answer was thought to be 'no', but Kuhl's recent studies have hinted that babies' ears can be kept open to other languages, given the right stimulation. When she exposed nine-month-old American babies to twelve twenty-five-minute language sessions over the course of a month in which a Mandarin speaker read from children's books and played with puppets, a train and stacks of rings, she found that these babies retained the ability to perceive Mandarin sounds when other American babies of the same age lost it. Importantly, this happened only if the babies were exposed to Mandarin speakers in the flesh – it didn't work if they were played a DVD in Mandarin. It's also unclear how long the effect lasts. In Kuhl's study, the babies retained their ear for Mandarin for at least two weeks after last hearing it but, unfortunately, they weren't tested again after that.

144

When is the best time to introduce a second language?

A KEY QUESTION that has troubled parents and linguists alike is whether children should master one language before attempting to tackle a second one. The fear has been that they may confuse the two languages or fall behind their peers who are just learning one. Laura-Ann Petitto of Gallaudet University in Washington DC has spent her career investigating how babies acquire language. She even spent some time trying to teach a chimpanzee to learn sign language.

Her research suggests this fear is unfounded. She tested

children who had been exposed to both French and English or French and sign language from birth and compared the age at which they reached language milestones such as speaking their first word or stringing words together in a sentence with children learning just one language. The bilingual children achieved these milestones at the same age, despite having spent only half as much time being exposed to either language.

'Considerable research now suggests that learning two languages is as natural as learning one and that, given the right learning environment, most children can acquire two languages simultaneously at the same rate and in the same way as monolingual children,' says Fred Genesee, a bilingualism expert at McGill University.

Another myth is the 'one parent, one language rule', or bilingual parents sticking to a single language when in the company of their child to avoid it becoming confused. 'There's absolutely no evidence that that's true,' says Genesee. 'However, it's not a bad strategy to use because it guarantees at least a minimum exposure to the minority language.'

If a baby lives in a house where more than one language is spoken, or if its mother speaks a different language to the community she's living in, it probably starts learning aspects of those languages even before birth. 'Starting at that point makes sense for children for whom being bilingual is really important, or who are likely to keep on being exposed to a language once they're born,' says Genesee. 'There's a lot of evidence suggesting that the first year or two establishes some critical foundations for language learning.'

What about monolingual parents who would simply like their child to be able to speak more than one language? Although there's no evidence that exposing such babies to more than one language is harmful, it could be a waste of time, because they're unlikely to experience enough of it to learn very much. Research suggests that children need to spend between 30 and 50 per cent of their day hearing a particular language in order to become as good at it as a monolingual child.

This suggests that my mother-in-law's attempts to teach our children French when she sees them every six weeks or so are unlikely to result in any degree of fluency, although it may help tune their ears to French, which could have benefits if they choose to study it when they're older.

In order to become proficient in a language, young children typically need two to three years of exposure to it. Immersing a young child in a foreign language for a few months would probably be enough to give them a good grounding in it, particularly if they are interacting with other children, but Genesee cautions against trying to bombard children with too many languages, as could happen if parents were regularly moving from country to country. 'What could be detrimental is if children are exposed to more than two or three languages and that exposure is fragmented and not extensive enough,' he says.

The Next One

145

Are women more fertile after having a baby?

PARENTING CHAT ROOMS are filled with anecdotal reports that women are super-fertile after having had a baby. Similar rumours abound about miscarriage, with many women claiming their doctor has told them they're more likely to conceive after a miscarriage because their hormone levels are more favourable.

Surprisingly, there's no research either to support or refute reports of extreme fertility in women who have already conceived a child. While some women appear to have been told that high progesterone levels following a miscarriage or pregnancy make them more likely to conceive, there is no published evidence to suggest that this is true.

Progesterone is a hormone that is produced in large amounts during pregnancy as well as at certain times during the menstrual cycle. It causes the lining of the uterus – the endometrium – to thicken, which could, theoretically, help a fertilized egg to implant. Indeed, some researchers are experimenting with progesterone to improve success rates during fertility treatment. However, 'It is not likely that progesterone remains at a high enough concentration to have an effect the month after a miscarriage,' says Anneli Stavreus Evers, a reproductive endocrinologist at Uppsala University Hospital in Sweden, who is leading some of these efforts. After a miscarriage, the lining of the uterus will be shed, just like during a normal period, and after that the lining should just develop in the same way as during a normal menstrual cycle. What's more, high progesterone at the wrong time of the menstrual cycle might actually impede implantation.

Another possibility is that previous pregnancies might stimulate improvements in medical conditions associated with reduced fertility. For example, changes in hormone levels during pregnancy could cause endometriosis, a common condition in which cells from the lining of the uterus grow elsewhere, to regress, while the growing uterus could also cause scar tissue within the uterus (caused by previous infections, or surgery) to dissolve, says Calleb Kallen, a reproductive endocrinologist and fertility expert at Emory University School of Medicine in Atlanta, Georgia. Some women who had infrequent periods before pregnancy find that their periods become more regular after having a baby. However, such fertility changes have never been systematically studied and are unlikely to affect more than a small minority of women. 'I am not aware of any genuine fertility or hormonal benefits stemming from delivering a baby,' says Kallen.

146

Am I likely to fall pregnant when breastfeeding?

YOUR CHANCES OF getting pregnant are greatly reduced, but it's no guarantee. Several studies have shown that if a woman suckles her baby more than five times a day for at least an hour in total, then she is highly unlikely to get pregnant – assuming the baby is latching on properly and feeding. The odds increase slightly after about six months, when babies start eating solids, but women who continue to breastfeed and aren't having periods are still unlikely to conceive.

The main trigger for this temporary infertility is the physical act of sucking the nipple. The maturation of eggs in the ovaries is controlled by a delicate balance of hormones that act as a kind of clock. The brain usually produces regular pulses of a hormone called gonadotrophin-releasing hormone (GnRH), which in turn causes another hormone called lutenizing hormone (LH) to be released, and this is what causes eggs to mature. The timing of these pulses is essential: GnRH needs to be released around once an hour for the right dose of LH to be secreted. Suckling slows the timing of these pulses, and if it slows too much then the result is temporary infertility.

When women stop breastfeeding, the pulses of GnRH begin to speed up again, but the rate at which this happens isn't the same for everybody. Some women who suddenly stop breastfeeding will fall pregnant without even having a period, but for others it takes time for the balance of these hormones to re-establish themselves. This is especially true if you stop breastfeeding during the first six months of having a baby. 'Even in the absence of breastfeeding the return of ovulation is delayed,' says Ronald Gray of Johns Hopkins Bloomberg School of Public Health in Baltimore. Indeed, many of the first bleeds that women experience after having a baby don't involve an egg being released.

To confuse things still further, mothers who slowly cut down their breastfeeding may find that their periods resume, even though they aren't ovulating and remain temporarily infertile. This is because pulses of GnRH have increased enough for the ovaries to start producing hormones that cause the uterine lining to thicken, but the signal isn't strong enough to mature an egg. When no egg is released, these hormone levels fall and women experience a period, even though they haven't ovulated.

147

I had no problem getting pregnant last time, so why am I struggling to conceive now?

WHILE SOME WOMEN fall pregnant at the first attempt, others struggle to conceive a second child, despite having found it relatively easy to get pregnant the first time around. Such 'secondary infertility' can be tough to bear, particularly if well-meaning friends and relatives are constantly asking when the first child is going to get a sibling.

According to the US National Infertility Association, Resolve, approximately 30 per cent of couples seeking help with infertility already have at least one child, and secondary infertility seems to be on the rise – up from around 1.8 million cases per year in the

US in 1995 to nearly 3 million in 2006. Despite this, there is little research into the causes of secondary infertility, as it is assumed that the causes are the same as those affecting couples who have never had a baby.

There are, however, some known mechanisms by which having a child might reduce subsequent fertility. The most obvious is breastfeeding, which vastly reduces a woman's chances of ovulating and becoming pregnant, at least during the first six months. Even if your periods have resumed, this doesn't necessarily mean that you are ovulating, as it can take time for the hormones that drive egg production in the ovaries to rebalance themselves (see 146: 'Am I likely to fall pregnant when breastfeeding?').

Ovulation might also be delayed if the body isn't getting enough energy to supply its needs. Claudia Valeggia at the University of Pennsylvania in Philadelphia recently measured hormone levels in seventy breastfeeding mothers in Toba, Argentina, and found that the average length of infertility after giving birth was 4.3 months. Closer analysis revealed that levels of insulin, which is produced in response to food, were low during the early months. This could be a sign that the women's energy supplies were in the red, possibly because the body needs a lot of energy to rebuild itself after pregnancy and birth.

Insulin helps the ovaries to produce reproductive hormones such as oestrogen and progesterone, so its availability may influence when women become fertile again. Shortly before the Toba women resumed ovulation they experienced a surge of insulin and tended to gain a few pounds in weight. 'Energy balance has a deep influence on the reproductive axis of women of all ages, so if you are trying to lose those "baby kilos" after your pregnancy, that may influence your chances of conceiving again,' says Valeggia. Carrying too many extra pounds may also interfere with how your body responds to insulin, while the stress of trying to juggle a career while running around after a toddler may dent your fertility still further.

In a minority of women, pregnancy can affect the function of the thyroid gland, resulting in something called postpartum thyroiditis. Around 7 per cent of women are estimated to develop this condition within the first year of giving birth, and common symptoms include hair loss, tiredness and depression. While all of these may be symptoms of new motherhood, thyroid function can affect your fertility, so it may be another thing to get checked out.

Some doctors believe that those suffering from secondary infertility may have always had an underlying problem, such as polycystic ovaries, and they simply got lucky the first time around. Fibroids also have a tendency to grow during pregnancy, so if they were already present they could now be having a greater impact on your fertility.

However, possibly the main culprit of secondary infertility is ageing. There is a sharp decline in fertility after the age of thirty-five, so it can simply take longer to fall pregnant. 'That said, prior fertility success means that, at least historically, everything worked at least once. So there must have been an open tube, a functional uterus, an egg and a sperm,' says Kallen.

If you are over the age of thirty-five and have been trying to conceive for more than six months, you should seek medical advice. Tests can be done to determine whether you are ovulating and if your levels of thyroid hormones and insulin are normal. If you are under thirty-five, the current advice is to wait a year before seeking help. Disconcerting as secondary infertility may be, it is quite normal to take a year or two to conceive – regardless of how quickly you got pregnant the first time around.

148

Why do some women seem to give birth only to boys?

On 17 January 1995, an earthquake measuring 6.8 on the Richter scale hit Kobe in Japan, killing around 6,434 people, injuring many more, and leaving thousands homeless. Nine months later, the first babies who would have been conceived during that traumatic time were born. Whereas, normally, around 51.6 per cent of babies born in Japan are male, in the Kobe area this fell to 50.1 per cent in the aftermath of the earthquake. Similar declines in male births have been seen following other disasters, including the collapse of the Twin Towers on 9/11, while an increase in male births was observed after the First and Second World Wars.

The received wisdom is that we all have an equal chance of having a son or daughter – that it's like flipping a coin – heads or tails – and families with 'runs' of boys or girls are just the product of coincidence. In fact, the coin is already weighted slightly in favour of boys, with approximately 106 boys born for every 100 girls worldwide (although this figure has fluctuated over the last century). No one really understands why this is (and the numbers eventually balance out, as males are more likely to die during childhood or before they have children), but we do know of several environmental factors that seem to influence the gender of children to some degree.

As a general rule, men produce approximately 50 per cent 'male' and 50 per cent 'female' sperm, but exposure to radiation or high levels of hormone-disrupting chemicals such as phthalates (which are added to plastics) seems to tip the balance in favour of 'female' sperm carrying an X-chromosome. For example, twice as many girls as boys are being born in a Canadian First Nations tribe called the Aamjiwnaang, which lives next door to the Sarnia-Lambton Chemical Valley complex in Ontario. High levels of hexachlorobenzene (HCB) have been found in the local soil, while it is also thought that phthalates are also being emitted from the complex, prompting some to speculate that these are skewing the sex ratio. Meanwhile, men working at the Sellafield nuclear plant in Cumbria, UK, are up to 40 per cent more likely to have boys than girls, suggesting that small doses of radiation may also have an effect on the sex ratio.

Exposure to high levels of chemicals and radiation is, admittedly, rare, but it does at least suggest that the possibility to skew sex ratios exists. More prosaically, recent research suggests that women working in high-stress jobs are more likely to give birth to daughters than sons. However, the effect was small: 54 per cent of women in low-stress jobs had sons, compared with 47 per cent of women in high-stress jobs. It also seemed to be influenced by how much money the father made, with high-salary dads slightly more likely to have sons.

Other studies have also found that male managers have more sons than their employees, as do men working in typically male occupations such as engineering. There are also certain diseases that seem to cause women to produce more sons, such as polycystic ovarian syndrome and multiple sclerosis.

How could all of this be? From an evolutionary perspective, it makes more sense to have boys when times are good – they take slightly more energy to grow and prosper, but when they do, they have the potential to generate far more grandchildren than do daughters (at least, they did in our ancestors' times), and so ensure the propagation of our genes. In lean times, daughters are a better guarantee of producing at least a couple of grandchildren, since men are less picky about with whom they mate (or so the theory goes).

Times of extreme stress may prompt the release of hormones that tell the body that lean times are ahead, either killing off male sperm or making male embryos (which tend to be slightly more fragile) less likely to survive. Similar hormonal factors may also be at work in the more everyday cases of occupation-related imbalances in the sex ratio.

It's also possible that some men or women are just inherently more likely to have boys than girls, although precisely why remains a mystery. In a study of 1,403,021 children born to 700,030 couples in Denmark, approximately 51.2 per cent of first births were male. But, as couples went on to have more children, an interesting pattern began to emerge. Women who had boys in their first and second births seemed to keep on having boys. In couples who already had three boys, the chances of having a fourth boy were 52.4 per cent, and this rose to 54.2 per cent for a fifth child. No such pattern was seen in couples with girls. So it seems that once you start down the path of having boys, you're statistically more likely to keep on having them.

A similar mechanism may lie behind the finding that men with lots of brothers are more likely to have sons, while those with many sisters tend to have daughters. Corry Gellatly at Newcastle University uncovered this trend after analysing 927 family trees containing information on 556,387 people from Europe and North America stretching back to the year 1600. Once again, it seems to be only men who are affected by this, so if a woman has lots of sisters, it is irrelevant. Gellatly proposes that some as yet undiscovered gene can influence the proportion of X- and Y-bearing sperm, which would mean that the assumption that all men have equal numbers of X- and Y-carrying sperm is untrue.

Age and beauty also enter into the equation. Several studies have found that the older you get, the less likely you are to have a son. One recent study found that 53 per cent of teenage parents have sons, while among parents over forty the proportion is just 35 per cent. Satoshi Kanazawa of the London School of Economics reached this conclusion after analysing data from the UK National Child Development study, which followed a population of 17,419 British-born children through their lives. Here, the effect seemed particularly pronounced among older mothers. Kanazawa calculates that for every year a woman ages, the odds of her first child being a boy decrease by 1.2 per cent. Egg quality is an obvious suspect: perhaps more fragile male foetuses are less able to survive with minor mutations in their DNA than females are.

Whatever the mechanism, Kanazawa believes there is a good evolutionary explanation for older parents having daughters rather than sons. Once conceived, there is little that parents can do to increase their daughters' future reproductive success beyond keeping them alive and healthy. Sons, on the other hand, traditionally needed more investment from their parents in order to make sure they inherited the family's status and resources. If a son's parents die before he reaches maturity, he is unlikely to be powerful enough to defend the family's wealth and resources. And although an orphaned girl is also in a perilous predicament, she may yet find a husband to take care of her. (We're talking in evolutionary terms here, not about modern-day society.) It therefore makes sense for older parents to have girls, Kanazawa says.

A friend of mine with two teenage daughters provided an alternative explanation: 'Mothers need all the age and experience they can get to handle teenage girls, so best to have them as late as possible,' she says.

In addition, Kanazawa found that beautiful parents were statistically more likely to have daughters than unattractive parents. When children in the study were seven, their teachers were asked to rate them as 'attractive', 'unattractive', 'looks scruffy and dirty', 'underfed' or 'abnormal features'. When those children grew up, the proportion of sons among those rated attractive was 50 per cent, compared to 52 per cent among those who fell into the other physical categories.

Again, Kanazawa has an evolutionary explanation. Although physical attractiveness is important for both sexes, it is especially important for women. Men prefer to mate with attractive women regardless of whether it's a one-night stand or a long-term partnership, while other traits such as wealth and status may be more important for women seeking a long-term partner.

It's important to bear in mind that, with all these studies, the effect on the sex ratio is small, which means that many ugly, young, chilled-out wives of business executives will give birth to daughters, and many stressed, ageing supermodels will have sons. But it's food for thought. The more people look into it, the less certain the assumption that gender is a pure matter of chance seems to be.

149

Is there anything I can do to influence the gender of my baby?

THE ONLY WAY to guarantee the gender of your baby is to undergo clinical sex selection (which is illegal in most countries), which involves IVF. But there are several 'natural' methods that may tip the balance slightly in your favour.

All babies carry two sex chromosomes (one from the mother and one from the father), which dictate their gender. Male babies carry an X and a Y chromosome, while female babies carry two X chromosomes. Since all eggs carry an X chromosome, it's the father's sperm that determines the sex of the baby. Men usually produce similar numbers of sperm bearing X and Y chromosomes so it's largely pot luck whether you end up with a baby boy or girl.

That's not to say, however, that external factors don't have any influence. The most famous method for influencing your baby's gender is the Shettles method, which is based on the premise that the X chromosome is larger than the Y chromosome, so it weighs slightly more and may slow 'female' sperm down. For this reason, couples wanting boys are told to have sex as close to the moment of ovulation as possible in the hope that male sperm will beat female sperm to the egg. However, female sperm are supposedly more robust, which means that they can survive for longer in the harsh environment of the reproductive tract. For this reason, Shettles advises couples wanting a girl to have sex two to three days before ovulation and to aim for shallower penetration to increase the amount of time the sperm spend in the acidic environment of the vagina.

Several studies have set out to test the Shettles method, but none has conclusively proved that it works. A key problem is that couples trying to conceive rarely have sex only once around the time of ovulation, and it's also tricky to pinpoint the exact moment of ovulation. At most, you might increase your chances of having the boy or girl you want by a couple of per cent, but that's nowhere near a guarantee. Plus, if you restrict the days on which you have sex, you're less likely to get pregnant in the first place.

Another common belief is that a woman's diet in the run-up to pregnancy can influence the sex of a baby. A survey of 740 pregnant women found that 56 per cent of women who had a high-energy diet before getting pregnant had boys, compared with 46 per cent who ate low-calorie diets.

One possibility is that women who eat more have higher blood-sugar levels, and this favours the survival of male embryos – something that has also been noticed when embryos are grown in IVF clinics. Women who gain weight between their first and second pregnancies also seem more prone to having boys the second time around, whereas women with anorexia nervosa seem more likely to have girls.

This also makes sense from an evolutionary perspective. Sons need slightly more energy to grow and tend to be bigger when they're born. They are also less likely to survive if they are malnourished as children, and smaller men tend to be less attractive as potential mates. So, in times of plenty, it makes sense to produce sons, as, if they survive, they have the potential to sire more grandchildren. But when food is scarce, hardier daughters may be a better option.

The mineral content of the women's diet may also have had an effect. Several studies have shown that women who eat diets rich in salt and potassium (found in bananas and spinach, among other things) are more likely to have boys, while those eating lots of calcium (found in milk and dairy products) and magnesium (found in nuts, seeds and dark chocolate) are more likely to have girls, although the reason for this is unclear.

You could also try combining these methods. In a recent study, twenty-one women who wanted to have girls were given kits to help them predict when they were ovulating and were told to stop having sex two days before they were due to ovulate. They also ate diets high in calcium and magnesium and low in sodium and potassium for nine weeks before trying to conceive. Sixteen of the women gave birth to a daughter, suggesting that the method may have worked, although the low number of women in the study means that it could simply have been down to chance. Other studies have found no effect.

Even if these methods work at a population level, there's no guarantee they will work for you – they may simply tip the balance slightly in your favour. So, on an individual level, you may just have to get on with it and be happy with what you get.

150
Do twins run in families?

SOME PEOPLE MELT at the prospect of two identical babies who will always share a special bond, while others quake in horror, but, if you've dismissed the prospect of twins because no one else in your family has had them, think again.

There are two types of twins: identical (monozygotic), which account for around a third of all twins, and non-identical (dizygotic), which account for the remaining two-thirds. Although non-identical twins do seem to run in families to some extent, identical twins (which are caused by an embryo splitting soon after fertilization) seem to be a completely random event.

Despite the name, identical twins aren't actually identical at all. Although they start out with the same genes, mutations creep in during the many rounds of cell division that enable a baby to grow, which explains why sometimes one twin develops a genetic disease when the other doesn't. Precisely where in the uterus the embryos implant can also influence how much blood and how many toxins twins are exposed to, which can lead to differences in growth.

Non-identical twins seem to run on the mother's side of the family, so if your mum or your sister had twins, you are twice as likely to have them as the average woman, possibly because there is some hereditary factor that causes more than one egg to be released each month at ovulation. The risk of having non-identical twins quadruples with age, as the pool of remaining eggs begins to dwindle and the body produces more of the hormone that stimulates the ovaries to release eggs.

Although the number of identical twins born is relatively consistent around the world, the rate of dizygotic twinning varies widely, presumably because of genetic factors. In Nigeria, around one in every eleven newborns is a twin, while in Japan it is more like one in every two hundred and fifty. In Europe and the US, between ten and twenty in every thousand births results in a twin.

Identical twins seem to be more common in women undergoing IVF, even if they have only one embryo transferred. This may have something to do with the physical processes involved in IVF or the way embryos are cultured before implantation. One recent study hinted that identical twins were more likely to form if there were structural weaknesses in the very early embryo.

This actually happened to one of my colleagues, who already had one child through IVF and hoped to have a second by implanting one of the remaining embryos. She was shown a photo of her defrosted embryo shortly before it was implanted then had a scan four weeks later, only to be confronted with two embryos sitting side by side in her uterus. Just imagine being able to sit with your twin and see a photo of yourself when you were just one individual, rather than two.

Notes

ACOG: American College of Obstetricians and Gynecologists
AJOG: *American Journal of Obstetrics and Gynecology*
BJOG: *British Journal of Gynaecology: An International Journal of*
          Obstetrics and Gynaecology
BMJ: *British Medical Journal*
DOI: Department of Information

FOREWORD

'Sense about science and straight statistics. Making sense of statistics',
http://www.senseaboutscience.org/data/files/resources/1/
MSofStatistics.pdf

ONE: BUMP

Food and Drink

Why do pregnant women crave unhealthy food?
http://www.babycenter.com/0_food-cravings-and-what-they-
mean_1313971.bc

'A Brief Review on How Pregnancy and Sex Hormones Interfere
with Taste and Food Intake', *Chemosensory Perception*, 2010, vol. 3 (1),
pp. 51–6

'Impact of the menstrual cycle on determinants of energy balance: a
putative role in weight-loss attempts', *International Journal of Obesity*,
2007, vol. 31, pp. 1777–85

'Disguised protein in lunch after low-protein breakfast conditions food-flavour preferences dependent on recent lack of protein intake', *Physiology and Behaviour*, 1995, vol. 58 (2), pp. 363–71
http://news.bbc.co.uk/1/hi/health/7370524.stm

Do pregnant women really eat coal?
'Pica in pregnancy: new ideas about an old condition', *Annual Review of Nutrition*, 2010, vol. 30, pp. 403–22

What causes morning sickness?
http://humupd.oxfordjournals.org/content/11/5/527.long
'Maternal serum HCG is higher in the presence of a female fetus as early as week three post-fertilization', *Human Reproduction*, 2002, vol. 17 (2), pp. 485–9
'Morning sickness: a mechanism for protecting mother and embryo', *The Quarterly Review of Biology*, 2000, vol. 77 (2), pp. 113–48
A Visitor Within: The Science of Pregnancy, David Bainbridge, Weidenfeld & Nicolson, 2000

Is there anything I can do to reduce morning sickness?
'Antenatal care: routine care for the healthy pregnant woman', NICE guidelines, National Institute for Health and Clinical Excellence, 2008
'ACOG Practice Bulletin: Nausea and vomiting of pregnancy', *Obstetrics and Gynecology*, April 2004, vol. 103 (4), pp. 803–14
'Interventions for nausea and vomiting in early pregnancy', Matthews, A., Dowswell, T., Haas, D. M., Doyle, M. and O'Mathúna, D. P., Cochrane Database of Systematic Reviews, 2010, issue 9, art. no. CD007575

How much alcohol is it safe to drink during pregnancy?
'Systematic review of effects of low–moderate prenatal alcohol exposure on pregnancy outcome', *BJOG*, March 2007, vol. 114 (3), pp. 243–52
http://www.rcog.org.uk/womens-health/clinical-guidance/alcohol-and-pregnancy-information-you
'Light drinking in pregnancy, a risk for behavioural problems and cognitive deficits at 3 years of age?' *International Journal of Epidemiology*, 2009, vol. 38 (1), pp. 129–40

'The effect of different alcohol drinking patterns in early to mid-pregnancy on the child's intelligence, attention, and executive function', *BJOG*, Sept. 2012, vol. 119 (10), pp. 1180–90

'"You can't *handle* the truth"; medical paternalism and prenatal alcohol use', *Journal of Medical Ethics*, 2009, vol. 35, pp. 300–3

'Prenatal alcohol exposure patterns and alcohol-related birth defects and growth deficiencies: a prospective study', *Alcoholism: Clinical and Experimental Research*, April 2012, vol. 36 (4), pp. 670–6

Can unborn babies taste what Mum is eating?

'Garlic ingestion by pregnant women alters the odor of amniotic fluid', *Chemical Senses*, April 1995, 20 (2), pp. 207–9

'Flavor perception in human infants: development and functional significance', *Digestion*, 2011; 83 Suppl. 1, pp. 1–6

'Prenatal and Postnatal Flavor Learning by Human Infants', *Pediatrics*, 1 June 2001, vol. 107, e88

Can Mum's food fads influence her baby's palate?

'Fetal Learning About Ethanol and Later Ethanol Responsiveness: Evidence Against "Safe" Amounts of Prenatal Exposure', *Experimental Biology and Medicine*, Feb. 2008, vol. 233 (2), pp. 139–54

'The interaction of gestational and postnatal ethanol experience on the adolescent and adult odor-mediated responses to ethanol in observer and demonstrator rats', *Alcohol Clinical and Experimental Research*, Oct. 2010, 34 (10), pp. 1705–13

'Infant salt preference and mothers' morning sickness', *Appetite*, June 1998, 30 (3), pp. 297–307

Is coffee bad for my baby?

'Effects of restricted caffeine intake by mother on fetal, neonatal and pregnancy outcome', Cochrane Database of Systematic Reviews, 15 April 2009, issue 2, art. no. CD006965

'Maternal caffeine intake from coffee and tea, fetal growth, and the risks of adverse birth outcomes: the Generation R Study', *American Journal of Clinical Nutrition*, June 2010, vol. 91 (6), pp. 1691–8

'Espresso coffees, caffeine and chlorogenic acid intake: potential health implications', *Food and Function*, 2012, vol. 3, pp. 30–3

'Eating while you are pregnant', UK Food Standards Agency

Can I eat peanuts during pregnancy?

'Maternal consumption of peanut during pregnancy is associated with peanut sensitization in atopic infants', *Journal of Allergy and Clinical Immunology*, 1 Dec. 2010, 126 (6), pp. 1191–7

'Statement on the review of the 1998 COT recommendations of peanut avoidance', Committee on Toxicology of Chemicals in Food, Consumer Products and the Environment

'Effects of early nutritional interventions on the development of atopic disease in infants and children: the role of maternal dietary restriction, breastfeeding, timing of introduction of complementary foods, and hydrolyzed formulas', *Pediatrics*, 1 Jan. 2008, vol. 121 (1), pp. 183–91

Should pregnant women really eat for two?

'Despite 2009 guidelines, few women report being counselled correctly about weight gain during pregnancy', *AJOG*, 2011, vol. 205 (333), e1–6

'Gestational weight gain and long-term postpartum weight retention: a meta-analysis', *American Journal of Clinical Nutrition*, Nov. 2011, 94 (5), pp. 1225–31

http://www.americanpregnancy.org/pregnancyhealth/eatingfortwo.html

How dangerous is it to eat Camembert and blue cheese?

http://www.nhs.uk/conditions/listeriosis/Pages/Introduction.aspx

'Listeriosis in human pregnancy: a systematic review', *Journal of Perinatal Medicine*, vol. 39 (3), pp. 227–36

'Pregnancy-associated listeriosis', *Epidemiology and Infection*, Oct. 2010, vol. 38 (10), pp. 1503–9

'Listeria monocytogenes infection from foods prepared in a commercial establishment: a case-control study of potential sources of sporadic illness in the United States', *Clinical Infectious Diseases*, 2007, vol. 44 (4), pp. 521–8

'Attribution of human listeria monocytogenes infections in England and Wales to ready-to-eat food sources placed on the market: adaptation of the Hald salmonella source attribution model', *Foodborne Pathogens and Disease*, 2010, vol. 7 (7), pp. 749–56

THE PREGNANT BODY

Can the shape of my bump or anything else predict the gender of my child?

'Food cravings, fetal heart rate, gut instinct . . .', *Birth*, vol. 26, p. 172

'Sickness in pregnancy and sex of child', *The Lancet*, vol. 354 (9195), p. 2053

'Hyperemesis gravidarum and sex of child', *The Lancet*, vol. 355 (9201), p. 407

'Gender-Related Differences in Fetal Heart Rate during First Trimester', *Fetal Diagnosis and Therapy*, 2006, vol. 21, pp. 144–7

'Fetal heart rate patterns in term labor vary with sex, gestational age, epidural analgesia, and fetal weight', *AJOG*, Jan. 1999, vol. 180 (1), pp. 181–7

'Average energy intake among pregnant women carrying a boy compared with a girl', *BMJ*, 7 June 2003, vol. 326 (7401), pp. 1245–6

'The relationship between placental location and fetal gender (Ramzi's method)', http://hcp.obgyn.net/ultrasound/content/article/1760982/1878451

'Placenta previa: preponderance of male sex at birth', *American Journal of Epidemiology*, 1999, vol. 149 (9), pp. 824–30

'The role of placental location assessment in the prediction of fetal gender', *Ultrasound in Obstetrics and Gynecology*, vol. 36, issue S1

Why don't pregnant women topple over?

'Fetal load and the evolution of lumbar lordosis in bipedal hominins', *Nature*, 13 Dec. 2007, vol. 450, p. 1075

Why do women get a linea nigra and other brown patches on their skin during pregnancy?

'Physiological and biological skin changes in pregnancy', *Clinics in Dermatology*, March–April 2006, vol. 24 (2), pp. 80–3

'Physiological changes in the skin during pregnancy', *Clinics in Dermatology*, Jan.–Feb. 1997, vol. 15 (1), pp. 35–43

Is there anything I can do to prevent stretch marks?

'Association of serum relaxin with striae gravidarum in pregnant women', *Archives of Gynecology and Obstetrics*, Feb. 2011, vol. 283 (2), pp. 219–22

'Effects of olive oil on striae gravidarum in the second trimester of pregnancy', *Complementary Therapies in Clinical Practice*, Aug. 2011, vol. 17 (3), pp. 167–9

'Prevention of striae gravidarum with cocoa butter cream', *International Journal of Gynaecology and Obstetrics*, Jan. 2010, vol. 108 (1), pp. 65–8

'Cocoa-butter lotion for prevention of striae gravidarum: a double-blind, randomized and placebo-controlled trial', *BJOG*, Aug. 2008, vol. 115 (9), pp. 1138–42

'Creams for preventing stretch marks in pregnancy', Young, G. and Jewell, D., Cochrane Database of Systematic Reviews, 1996, issue 1, art. no.: CD000066

Do big parents have bigger babies?

'Fetal growth and growth restriction', Peregrine, Elisabeth and Peebles, Donald, *Fetal Medicine: Basic Science and Clinical Practice*, Churchill, 2008

'Maternal predictors of birthweight: the importance of weight gain during pregnancy', *Obesity Research and Clinical Practice*, Dec. 2007, vol. 1 (4), pp. 243–52

Does a woman's body shape permanently change after pregnancy?

'Changes in body-fat distribution in relation to parity in American women: a covert form of maternal depletion', *American Journal of Physical Anthropology*, 2006, vol. 131, pp. 295–302

Can jacuzzis and saunas really cause miscarriage?

'Risks of hyperthermia associated with hot tub or spa use by pregnant women', *Birth Defects Research Part A: Clinical and Molecular Teratology*, Aug. 2006, vol. 76 (8), pp. 569–73

http://www.americanpregnancy.org/pregnancyhealth/hottubs.htm

Why do so many human embryos miscarry before twelve weeks?

'Chromosome instability is common in human cleavage-stage embryos', *Nature Medicine*, 2009, vol. 15, pp. 577–83

'Incidence of early loss of pregnancy', *New England Journal of Medicine*, 28 July 1988, vol. 319 (4), pp. 189–94

http://www.rcog.org.uk/womens-health/clinical-guidance/early-miscarriage-information-you

'Influence of past reproductive performance on risk of spontaneous abortion', *BMJ*, 1989, vol. 229, pp. 545–9

http://www.ivfauthority.com/2011/02/ivf-commonly-asked-questions-fears-and_09.html

EXERCISE

Is it safe to exercise during pregnancy?
'Aerobic exercise for women during pregnancy', Kramer, M. S. and McDonald, S. W., Cochrane Database of Systematic Reviews, 2006, issue 3, art. no.: CD000180

'Leisure time physical exercise during pregnancy and the risk of miscarriage: a study within the Danish National Birth Cohort', *BJOG*, 1 Nov. 2007, vol. 114 (11), pp. 1419–26

'Exercise in pregnant women and birth weight: a randomized controlled trial', *BMC Pregnancy and Childbirth*, 2011, vol. 11, p. 66

'Exercise during pregnancy. A clinical update', *Clinical Sports Medicine*, April 2000, vol. 19 (2), pp. 273–86

'Aerobic exercise during pregnancy influences fetal cardiac autonomic control of heart rate and heart rate variability', *Early Human Development*, April 2010, vol. 86 (4), pp. 213–17

'Strenuous exercise in pregnancy: is there a limit?', *AJOG*, Jan. 2012, vol. 206 (1), supplement, p. S71

'A prospective investigation of the US public health recommendations for physical activity during pregnancy', *AJOG*, Jan. 2012, vol. 206 (1), supplement, S71–S72

What stomach exercises can I do when pregnant?
Mayes' Midwifery. A Textbook for Midwives, 13th edn, Baillière Tindall, 2004

Will lying on my back harm my baby?
'Quantitative cardiovascular magnetic resonance in pregnant women: cross-sectional analysis of physiological parameters throughout pregnancy and the impact of the supine position', *Journal of Cardiovascular Magnetic Resonance*, 27 June 2011, vol. 13, p. 31

Do pelvic-floor exercises actually work?

'Pelvic floor muscle training for prevention and treatment of urinary and faecal incontinence in antenatal and postnatal women (Review)', Hay-Smith, J., Mørkved, S., Fairbrother, K. A. and Herbison, G. P., Cochrane Database of Systematic Reviews, 2008, issue 4, art. no.: CD007471

'Postpartum sexual function of women and the effects of early pelvic-floor-muscle exercises', *Acta Obstetrica et Gynecologica*, 2010, vol. 89, pp. 817–22

http://www.guysandstthomas.nhs.uk/resources/patientinfo/womens/pelvicfloorexercises.pdf

BABY ON THE BRAIN

Does pregnancy make women forgetful?

'Giving birth to a new brain: Hormone exposures of pregnancy influence human memory', *Psychoneuroendocrinology*, Sept. 2010, vol. 35 (8), pp. 1148–55

'Change in brain size during and after pregnancy: study in healthy women and women with preeclampsia', *American Journal of Neuroradiology*, 2002, vol. 23, pp. 19–26

'Motherhood and the hormones of pregnancy modify concentrations of hippocampal neuronal dendritic spines', *Hormones and Behavior*, Mar. 2006, vol. 49 (2), pp. 131–42

'Motherhood improves learning and memory', *Nature*, 11 November 1999, vol. 401, pp. 137–8

'Emotional sensitivity for motherhood: late pregnancy is associated with enhanced accuracy to encode emotional faces', *Hormones and Behavior*, Nov. 2009, vol. 56 (5), pp. 557–63

'Attenuation of maternal psychophysiological stress responses and the maternal cortisol awakening response over the course of human pregnancy', *Stress*, May 2010, vol. 13 (3), pp. 258–68

Is stress during pregnancy bad for my baby?

'Psychosocial stress and pregnancy outcome', *Clinical Obstetrics and Gynecology*, Jun. 2008, vol. 51 (2), pp. 333–48

'Psychosocial stress during pregnancy and perinatal outcomes: a meta-analytic review', *Journal of Psychosomatic Obstetrics and Gynaecology*, Dec. 2010, vol. 31 (4), pp. 219–28

'Prenatal Antecedents of Newborn Neurological Maturation', *Child Development*, Feb. 2010, vol. 81 (1), pp. 115–30

'Hormonal correlates of paternal responsiveness in new and expectant fathers', *Evolution of Human Behaviour*, 1 March 2000, vol. 21 (2), pp. 79–95

Can men get pregnancy symptoms too?
'A critical review of the Couvade syndrome: the pregnant male', *Journal of Reproductive and Infant Psychology*, Aug. 2007, vol. 25 (3), pp. 173–89

'Sad dads. Paternal postpartum depression', *Psychiatry*, Feb. 2007, vol. 4 (2), pp. 35–47

Why do women go into nesting overdrive in the final weeks of pregnancy?
'The plasticity of human maternal brain: longitudinal changes in brain anatomy during the early postpartum period', *Behavioural Neuroscience*, Oct. 2010, vol. 124 (5), pp. 695–700

'Breastfeeding, brain activation to own infant cry, and maternal sensitivity', *Journal of Child Psychology and Psychiatry, and Allied Disciplines*, Aug. 2011, vol. 52 (8), pp. 907–15

Are some women naturally more maternal than others?
'Specifying the neurobiological basis of human attachment: brain, hormones, and behavior in synchronous and intrusive mothers', *Neuropsychopharmacology*, Dec. 2011, vol. 36 (13), pp. 2603–15

Do men change when they become dads?
'Prolactin, Oxytocin, and the development of paternal behavior across the first six months of fatherhood', *Hormones and Behavior*, Aug. 2010, vol. 58 (3), pp. 513–8

'Natural variations in maternal and paternal care are associated with systematic changes in oxytocin following parent–infant contact', *Psychoneuroendocrinology*, Sept. 2010, vol. 35 (8), pp. 1133–41

'Variation in testosterone levels and male reproductive effort: Insight from a polygynous human population', *Hormones and Behavior*, Nov. 2009, vol. 56 (5), pp. 491–7

'Changes in testosterone, cortisol, and estradiol levels in men becoming fathers', Mayo Clinic Proceedings, June 2001, vol. 76 (6), pp. 582–92

Bumpology

THE DEVELOPING BABY

When does a baby become conscious?

'The emergence of the mind – a borderline of human viability?', *Acta Pediatrica*, 2007, vol. 96, pp. 327–8

'The emergence of human consciousness: from fetal to neonatal life', *Pediatric Research*, 2009, vol. 65 (3), pp. 255–60

'Fetal awareness and pain: what precautions should be taken to safeguard fetal welfare during experiments?', *Alternatives to Animal Testing and Experimentation*, vol. 14, special issue, pp. 79–83

Does a wriggly bump equal a boisterous baby?

'Optimization and initial experience of a multislice balanced steady-state free precession cine sequence for the assessment of fetal behavior in utero', *American Journal of Neuroradiology*, 2011, vol. 32, pp. 331–8

'Psychological and psychophysiological considerations regarding the maternal–fetal relationship', *Infant and Child Development*, Jan. 2010, vol. 19 (1), pp. 27–38

'What does fetal movement predict about behavior during the first two years of life?', *Developmental Psychobiology*, May 2002, vol. 40 (4), pp. 358–71

'Predicting infant crying from fetal movement data: an exploratory study', *Early Human Development*, 1999, vol. 54, pp. 55–62

'Twins' temperament: early prenatal sonographic assessment and postnatal correlation', *Journal of Perinatology*, May 2009, vol. 29 (5), pp. 337–42

Can a baby detect its mother's mood?

'Fetal response to induced maternal emotions', *The Journal of Physiological Sciences*, vol. 60 (3), pp. 213–20

'Prenatal origins of temperamental reactivity', *Early Human Development*, vol. 84 (9), Sept. 2008, pp. 569–75

What do babies learn during their time in the womb?

'Fetal psychology: an embryonic science', Hepper, Peter, in *Fetal Behaviour – Developmental and Perinatal Aspects*, J. Nijhuis (ed.), Oxford University Press, 1992

'Fetal memory: Does it exist? What does it do?', *Acta Paediatrica* supplement, vol. 416, pp. 16–20

'Aspects of fetal learning and memory', *Child Development*, 2009, vol. 80 (4), pp. 1251–8

Will playing Mozart to my bump make my baby more intelligent?

'Prelude to a musical life: prenatal music experiences', *Music Educators Journal*, 1985, vol. 71 (7), pp. 26–7

'Music and spatial task performance', *Nature*, 14 Oct. 1993, vol. 365 (6447), p. 611

'Improved maze learning through early music exposure in rats', *Neurological Research*, July 1998, vol. 20 (5), pp. 427–32

'Foetal response to music and voice', *Australia and New Zealand Journal of Obstetrics and Gynaecology*, Oct. 2005, vol. 45 (5), pp. 414–17

'Effect of music by Mozart on energy expenditure in growing preterm infants', *Pediatrics*, Jan. 2010, vol. 125 (1), pp. 24–8

Will my baby look like Mum or Dad?

'Genotype–phenotype associations and human eye color', *Journal of Human Genetics*, 2011, vol. 56, pp. 5–7

'Heritability of craniofacial characteristics between parents and offspring estimated from lateral cephalograms', *American Journal of Orthodontics and Dentofacial Orthopedics*, Feb. 2005, vol. 127 (2), pp. 200–7

'The similarity between parents and offspring in facial soft tissue: a frontal-view study', *Australian Orthodontic Journal*, March 1998, vol. 15 (2), pp. 85–92

'The use of parental data to evaluate soft tissues in an Anatolian Turkish population according to Holdaway soft tissue norms', *American Journal of Orthodontics and Dentofacial Orthopedics*, March 2006, vol. 129 (3), pp. 330.e1–330.e9

'Are political orientations genetically transmitted?', *American Political Science Review*, vol. 99 (2), pp. 153–67

Do unborn babies dream?

'Fetal Yawning', *Frontiers of Neurology and Neuroscience*, 2010, vol. 28, pp. 32–41

'REM sleep and dreaming: towards a theory of protoconsciousness', *Nature Reviews Neuroscience*, Nov. 2009, vol. 10, pp. 803–14

'Fetal neonatal sleep and circadian rhythms', *Sleep Medicine Reviews*, 2003, vol. 7 (4), pp. 321–34

'Circadian rhythms in the fetus', *Molecular and Cellular Endocrinology*, 2012, vol. 349, pp. 68–75

Are babies more active at night?
'Fetal neonatal sleep and circadian rhythms', *Sleep Medicine Reviews*, 2003, vol. 7 (4), pp. 321–34

How do fingerprints develop?
http://www.scientificamerican.com/article.cfm?id=are-ones-fingerprints-sim

'Fingerprint formation', *Journal of Theoretical Biology*, 7 July 2005, vol. 235 (1), pp. 71–83

'A fingerprint characteristic associated with the early prenatal environment', *American Journal of Human Biology*, 2008, vol. 20, pp. 59–65

'A fingerprint marker from early gestation associated with diabetes in middle age: the Dutch Hunger Winter families study', *International Journal of Epidemiology*, 2009, vol. 38, pp. 101–9

TWO: BIRTH

GET A MOVE ON

Who decides when it's time to come out?
A Visitor Within: The Science of Pregnancy, Bainbridge, David, Weidenfeld & Nicolson, 2000

'Endocrine immune interactions in human parturition', *Molecular and Cellular Endocrinology*, 15 March 2011, vol. 335 (1), pp. 52–9

'Are animal models relevant to key aspects of human parturition?', *American Journal of Psychiatry – Regulatory, Integrative and Comparative Physiology*, Sept. 2009, vol. 297 (3), R525–R545

Can my state of mind delay labour?
'Influence of Valentine's Day and Halloween on birth timing', *Social Science and Medicine*, vol. 73 (8), Oct. 2011, pp. 1246–8

Is it normal for pregnancies to run past their due date?
'Naegele's rule: a reappraisal', *BJOG*, Nov. 2000, vol. 107 (11), pp. 1433–5

'The length of uncomplicated human gestation', *Obstetrics and Gynecology*, June 1990, vol. 75 (6), pp. 929–32

'Timing of planned Cesarean delivery by racial group', *Obstetrics and Gynecology*, 2008, vol. 111 (3), pp. 659–66

'NCT evidence-based briefing: induction of labour', *New Digest*, Feb. 2002

Can curry or anything else help trigger labour?
'Complementary and alternative medicine for induction of labour', *Women and Birth*, Sept. 2012, vol. 25 (3), pp. 142–8

'Raspberry leaf – should it be recommended to pregnant women?', *Complementary Therapies in Clinical Practice*, 2009, vol. 15, pp. 204–8

Does having a membrane sweep work?
'Membrane sweeping for induction of labour', Boulvain, M., Stan, C. M. and Irion, O., Cochrane Database of Systematic Reviews, 2005, issue 1, art. no.: CD000451

Will being induced mean I'm less likely to have a natural birth?
'Prolonged pregnancy: evaluating gestation-specific risks of fetal and infant mortality', *BJOG*, Feb. 1998, vol. 105 (2), pp. 169–73

'Systematic review: elective induction of labor versus expectant management of pregnancy', *Annals of Internal Medicine*, 18 Aug. 2009, vol. 151 (4), pp. 252–63

'Labor induction versus expectant management for post-term pregnancies: a systematic review with meta-analysis', *Obstetrics and Gynecology*, June 2003, vol. 101 (6), pp. 1312–18

'Induction of labour for improving birth outcomes for women at or beyond term', Gülmezoglu, A. M., Crowther, C. A., Middleton, P., Cochrane Database of Systematic Reviews, 2006, issue 4, art. no. CD004945

'Comparison of induction of labour and expectant management in post-term pregnancy: a matched cohort study', *Journal of Midwifery and Women's Health*, Sept.–Oct. 2009, vol. 54 (5), pp. 351–6

'The effect of early oxytocin augmentation in labor: a meta-analysis', *Obstetrics and Gynecology*, Sept. 2009, vol. 114 (3), pp. 641–9

'High-dose vs low-dose oxytocin for labor augmentation: a systematic review', *AJOG*, Oct. 2010, vol. 203 (4), pp. 296–304; epub 8 May 2010

Will rocking on my hands and knees cause my baby to turn over?

'Changing fetal position through maternal posture', *Journal of Obstetrics and Gynaecology Research*, vol. 33 (3), pp. 279–82

'Changing fetal position through maternal posture', *Zhonghua Fu Chan Ke Za Zhi*, Sept. 1993, vol. 28 (9), pp. 517–19

'Hands and knees posture in late pregnancy or labour for fetal malposition (lateral or posterior)', Hunter, S., Hofmeyr, G. J. and Kulier, R., Cochrane Database of Systematic Reviews, 2007, issue 4, art. no.: CD001063

'Caesarean section', NICE Clinical Guideline (CG132), National Institute for Health and Clinical Excellence, Nov. 2011

Can I take a bath once my waters have broken?

'Warm tub bath during labor. A study of 1,385 women with prelabor rupture of the membranes after 34 weeks of gestation', *Acta Obstetrica et Gynecologica*, Aug. 1996, vol. 75 (7), pp. 642–4

THE BIG PUSH

Why do humans find it so difficult to give birth?

'Bipedalism and human birth: the obstetrical dilemma revisited', *Evolutionary Anthropology: Issues, News, and Reviews*, 2005, vol. 4 (5), pp. 161–8

Is there any way to predict how long labour will last?

'"Active labor". Duration and dilation rates among low-risk, nulliparous women with spontaneous labor onset: a systematic review', *Journal of Midwifery and Women's Health*, July–Aug. 2010, vol. 55 (4), pp. 308–18

'Racial differences in pelvic anatomy by magnetic resonance imaging', *Obstetrics and Gynecology*, 2008, vol. 111 (4), pp. 914–20

'The descent curve of the grand multiparous woman', *AJOG*, 2003, vol. 189 (4), pp. 1036–41

'Obese women have longer duration of first stage of labor', *AJOG*, 2012, vol. 206 (1), S149–50

'Maternal height, shoe size, and outcome of labour in white primigravidas: a prospective anthropometric study', *BMJ*, 20–27 Aug. 1988, vol. 297, pp. 515–17

'Identification of latent phase factors associated with active labor duration in low-risk nulliparous women with spontaneous contractions', *Acta Obstetrica et Gynecologica*, 2010, vol. 89, pp. 1034–9

Are vaginal births really better than C-sections?

Listening to Mothers II: Report of the Second National US Survey of Women's Childbearing Experiences, New York: Childbirth Connection, Oct. 2006, pp. 1–9

'Elective cesarean section and decision-making: a critical review of the literature', *Birth*, March 2007, vol. 34 (1), pp. 65–79

'Short-term maternal and neonatal outcomes by mode of delivery: a case-controlled study', *European Journal of Obstetrics and Gynecology and Reproductive Biology*, Nov. 2007, vol. 135 (1), pp. 35–40

A vaginal birth is harder on your sex life

'Postpartum sexual functioning and method of delivery: summary of the evidence', *Journal of Midwifery and Women's Health*, 2004, vol. 49 (5), pp. 430–6

'Effects of perineal trauma on postpartum sexual function', *Journal of Advanced Nursing*, 2010, vol. 66 (12), pp. 2640–9

'Women's sexual health after childbirth', *BJOG*, Feb. 2000, vol. 107 (2), pp. 186–95

'Which factors determine the sexual function one year after childbirth?', *BJOG*, Aug. 2006, vol. 113 (8), pp. 914–18

'Postpartum sexual functioning and its relationship to perineal trauma: a retrospective cohort study of primiparous women', *AJOG*, 2001, vol. 184 (5), pp. 881–90

Vaginal births increase the risk of incontinence

'Caesarean section', NICE Clinical Guideline (CG132), Nov. 2011

'Does cesarean section reduce postpartum urinary incontinence? A systematic review', *Birth*, Sept. 2007, vol. 34 (3), pp. 228–37

It takes longer to recover after a C-section
'Delivery method and self-reported postpartum general health status among primiparous women', *Paediatric and Perinatal Epidemiology*, 2001, vol. 15 (3), pp. 232–40
    'Caesarean section', NICE Clinical Guideline (CG132), Nov. 2011

C-sections make it harder to bond with your baby
'Women who have caesareans "less likely to bond"', *Daily Mail*, 13 July 2006
    'Initial handling of newborn infants by vaginally and cesarean-delivered mothers', *Nursing Research*, Sept.–Oct. 1986, vol. 35 (5), pp. 296–300
    'Maternal brain response to own baby-cry is affected by caesarean-section delivery', *Journal of Child Psychology and Psychiatry*, 2008, vol. 49 (10), pp. 1042–52
    'Breastfeeding, brain activation to own infant cry, and maternal sensitivity', *Journal of Child Psychology and Psychiatry*, Aug. 2011, vol. 52 (8), pp. 907–15
    'Sequelae of cesarean and vaginal deliveries: psychosocial outcomes for mothers and infants', *Developmental Psychology*, March 2000, vol. 36 (2), pp. 251–60
    'Mother and newborn baby: mutual regulation of physiology and behavior – a selective review', *Developmental Psychobiology*, Nov. 2005, vol. 47 (3), pp. 217–29

C-sections make breastfeeding more difficult
'Elective cesarean delivery: does it have a negative effect on breastfeeding?' *Birth*, Dec. 2010, vol. 37 (4), pp. 275–9
    'Early skin-to-skin after cesarean to improve breastfeeding', *MCN: The American Journal of Maternal Child Nursing*, Sept.–Oct. 2011, vol. 36 (5), pp. 318–24; quiz pp. 325–6

Babies born by C-section are less healthy
'Caesarean section', NICE Clinical Guideline (CG132), Nov. 2011
    'A meta-analysis of the association between caesarean section and childhood asthma', *Clinical and Experimental Allergy*, vol. 38, pp. 629–33

'Birth by caesarean section and asthma', *Clinical and Experimental Allergy*, vol. 38, pp. 554–6

'Caesarean delivery and risk of atopy and allergic disease: meta-analyses', *Clinical and Experimental Allergy*, April 2008, vol. 38 (4), pp. 634–42

Are home births more risky than hospital births?
'Home births', Royal College of Obstetricians and Gynaecologists/Royal College of Midwives joint statement no. 2, April 2007

'Intrapartum care. Care of healthy women and their babies during childbirth', National Collaborating Centre for Women's and Children's Health. Commissioned by the National Institute for Health and Clinical Excellence, Sept. 2007

'An estimation of intrapartum-related perinatal mortality rates for booked home births in England and Wales between 1994 and 2003', *BJOG*, April 2008, vol. 115 (5), pp. 554–9

'NHS choices: home birth risk remains unclear', 2 April 2008 (http://www.nhs.uk/news/2008/04April/Pages/Homebirthriskremainsunclear.aspx)

Does walking or squatting speed up labour?
'Maternal positions and mobility during first-stage labour', Lawrence, A., Lewis, L., Hofmeyr, G. J., Dowswell, T. and Styles, C., Cochrane Database of Systematic Reviews, 2009, issue 2, art. no.: CD003934

Should I push?
'A meta-analysis of passive descent versus immediate pushing in nulliparous women with epidural analgesia in the second stage of labor', *Journal of Obstetric, Gynecologic and Neonatal Nursing*, Jan.–Feb. 2008, vol. 37 (1), pp. 4–12

'Delayed versus immediate pushing in second stage of labor', *MCN: The American Journal of Maternal Child Nursing*, March–April 2010, vol. 35 (2), pp. 81–8

Is it better to tear or be cut in terms of healing?
'Episiotomy characteristics and risks for obstetric anal sphincter injuries: a case-control study', *BJOG*, May 2012, vol. 119 (6), pp. 724–30

Does perineal massage work?

'Antenatal perineal massage for reducing perineal trauma',
Beckmann, M. M. and Garrett, A. J., Cochrane Database of Systematic
Reviews, 2006, issue 1, art. no.: CD005123

'Antenatal perineal massage: information for women', Oxford Radcliffe
Hospitals NHS Trust

Is there anything else I can do to prevent tearing?

'Perineal techniques during the second stage of labour for reducing
perineal trauma', Aasheim, V., Nilsen, A. B. V., Lukasse, M. and Reinar,
L. M., Cochrane Database of Systematic Reviews, 2011, issue 12,
art. no.: CD006672

Will delaying cord-clamping benefit my baby?

'Effect of delayed versus early umbilical cord-clamping on neonatal
outcomes and iron status at four months: a randomized controlled trial',
BMJ, 2011, vol. 343, d7157

'Effect of timing of umbilical cord-clamping of term infants
on maternal and neonatal outcomes', McDonald, S. J. and
Middleton, P., Cochrane Database of Systematic Reviews, 2008,
issue 2. art. no.: CD004074

Ouch!

What's more painful: childbirth or having your leg chopped off?

'The nature of labour pain', *AJOG*, May 2002, vol. 186 (5), pp. S16–24

'More in hope than expectation: a systematic review of women's
expectations and experience of pain relief in labour', *BMC Medicine*,
Mar. 2008, vol. 6, art. 7

'Labour is still painful after prepared childbirth training', *CMA
Journal*, 15 Aug. 1981, vol. 125, p. 357

Do hormones block the pain of labour?

'The nature and consequences of childbirth pain', *European Journal of
Obstetrics and Gynecology and Reproductive Biology*, May 1995, vol. 59,
supplement, S9–S15

'A rise in pain threshold during labor: a prospective clinical trial',
Pain, Nov. 2007, vol. 132 (supplement 1), S104–8

'Antenatal women with or without pelvic pain can be characterized by generalized or segmental hypoalgesia in late pregnancy', *The Journal of Pain*, Dec. 2002, vol. 3 (6), pp. 451–60

Why do women come back for more?

'A longitudinal study of women's memory of labour pain', *BJOG*, Mar. 2009, vol. 116 (4), pp. 577–83

'Selective amnesic effects of oxytocin on human memory', *Physiology and Behavior*, Oct. 2004, vol. 83 (1), pp. 31–8

Can hypnosis or alternative therapies reduce labour pain?

'Pain management for women in labour: an overview of systematic reviews', Jones, L., Othman, M., Dowswell, T., Alfirevic, Z., Gates, S., Newburn, M., Jordan, S., Lavender, T. and Neilson, J. P., Cochrane Database of Systematic Reviews, 2012, issue 3, art. no.: CD009234

'Immersion in water in labour and birth', Cluett, E. R. and Burns, E., Cochrane Database of Systematic Reviews, 2009, issue 2, art. no.: CD000111

'Acupuncture for pain relief in labour: a systematic review and meta-analysis', *BJOG*, July 2010, vol. 117 (8), pp. 907–20

'Aromatherapy for pain management in labour', Smith, C. A., Collins, C. T. and Crowther, C. A., Cochrane Database of Systematic Reviews, 6 July 2011, issue 7, art. no.: CD009215

Does an epidural make a C-section more likely?

'Epidural versus non-epidural or no analgesia in labour', Anim-Somuah, M., Smyth, R. M. D. and Jones, L., Cochrane Database of Systematic Reviews, 2011, issue 12, art. no.: CD000331

'Pain management for women in labour: an overview of systematic reviews', Jones, L., Othman, M., Dowswell, T., Alfirevic, Z., Gates, S., Newburn, M., Jordan, S., Lavender, T. and Neilson, J. P., Cochrane Database of Systematic Reviews, 2012, issue 3, art. no.: CD009234

'Anesthesia and analgesia-related preferences and outcomes of women who have birth plans', *Journal of Midwifery and Women's Health*, July–Aug. 2011, vol. 56 (4), pp. 376–81

'Labour analgesia and the baby: good news is no news', *International Journal of Obstetric Anesthesia*, January 2011, vol. 20, issue 1, pp. 38–50

http://www.allaboutepidural.com/main-content/epidural-and-instrumental-delivery-forceps-and-vacuum

'Risk factors for perineal injury during delivery', *AJOG*, July 2003, vol. 189 (1), pp. 255–60

Do gas and air or pethidine relieve pain?

'Labour analgesia and the baby: good news is no news', *International Journal of Obstetric Anesthesia*, Jan. 2011, vol. 20 (1), pp. 38–50

'Pain management for women in labour: an overview of systematic reviews', Jones, L., Othman, M., Dowswell, T., Alfirevic, Z., Gates, S., Newburn, M., Jordan, S., Lavender, T. and Neilson, J. P., Cochrane Database of Systematic Reviews, 2012, issue 3, art. no.: CD009234

'A double-blinded randomised placebo-controlled study of intramuscular pethidine for pain relief in the first stage of labour', *BJOG*, July 2004, vol. 111 (7), pp. 648–55

THE POST-PREGNANCY BODY

What causes the baby blues?

'Elevated brain Monoamine Oxidase A binding in the early postpartum period', *Archives of General Psychiatry*, May 2010, vol. 67 (5), p. 468

'Evidence-based recommendations for depressive symptoms in postpartum women', *Journal of Obstetric, Gynecologic and Neonatal Nursing*, 11 March 2008, vol. 37 (2), p. 127

Why can't I poo?

Mayes' Midwifery. A Textbook for Midwives, 13th edn, Baillière Tindall, 2004

'Constipation', St Mark's Hospital, London, patient information leaflet

How do I know if my stitches are OK?

'Perineal care', *Clinical Evidence* (online), BMJ Publishing Group, March 2010

'Obstetric perineal wound infection: is there under-reporting?', *British Journal of Nursing*, 8 March 2012, vol. 21 (5), S28

'Minimizing postpartum pain: a review of research pertaining to perineal care in childbearing women', *Journal of Advanced Nursing*, August 2000, vol. 32 (2), p. 407

'Local cooling for relieving pain from perineal trauma sustained during childbirth', East, C. E., Begg, L., Henshall, N. E., Marchant, P. R. and Wallace, K., Cochrane Database of Systematic Reviews, 16 May 2012, issue 5, art. no.: CD006304

Is there anything I can do to stop my breasts sagging after breastfeeding?

'Breastfeeding and perceived changes in the appearance of the breasts: a retrospective study', *Acta Pediatrica*, 2004, vol. 93 (10), pp. 1346–8

'Breast ptosis: causes and cure', *Annals of Plastic Surgery*, May 2010, vol. 64 (5), pp. 579–84

'The female breast: structure and changes in pregnancy and lactation', *Midwives Information and Resource Service practice development*, Jun. 2011

Do women who breastfeed really lose their baby-weight faster than those who bottle-feed?

'Impact of breastfeeding on maternal nutritional status', *Advances in Experimental Medicine and Biology*, 2004, vol. 554, pp. 91–100

'Balancing exercise and food intake with lactation to promote post-partum weight loss', *The Proceedings of the Nutrition Society*, May 2011, vol. 70 (2), pp. 181–4

Do cells from my baby live on in my body after birth?

'Fetal microchimerism as an explanation of disease', *Nature Reviews Endocrinology*, Feb. 2011, vol. 7, pp. 89–97

'Fetal microchimerism in the maternal mouse brain: a novel population of fetal progenitor or stem cells able to cross the blood–brain barrier?', *Stem Cells*, Nov. 2005, vol. 23 (10), pp. 1443–52

'Fetal microchimerism: benevolence or malevolence for the mother?', *European Journal of Obstetrics, Gynecology and Reproductive Biology*, Oct. 2011, vol. 158 (2), pp. 148–52

'Pregnancy, microchimerism and the maternal grandmother', *PLoS One*, 2011, vol. 6 (8), e24101

THREE: BABIES

Portrait of a Newborn

Does birth distress the baby?

'Mode of delivery modulates physiological and behavioral responses to neonatal pain', *Journal of Perinatology*, Jan. 2009, vol. 29, pp. 4–50

'Newborn analgesia mediated by oxytocin during delivery', *Frontiers in Cellular Neuroscience*, Apr. 2011, vol. 5, art. 3

'Effects of oxytocin on GABA signalling in the foetal brain during delivery', *Progress in Brain Research*, 2008, vol. 170, pp. 243–57

Why do newborns look like their dads?

'Whose baby are you?' *Nature*, 14 Dec. 1995, 378 (6558), p. 669

'Measuring paternal discrepancy and its public health consequences', *Journal of Epidemiology and Community Health*, 2005, vol. 59, pp. 749–54

'The resemblance of one-year-old infants to their fathers: refuting Christenfeld and Hill (1995)', *Proceedings of the twenty-second annual conference of the Cognitive Science Society*, p. 148

What do newborns know?

Introduction to Infant Development, Alan Slater and Michael Lewis (eds), 2nd edn, 2007, Oxford University Press

'Of human bonding: newborns prefer their mothers' voices', *Science*, 1980, vol. 208 (4448), pp. 1174–6

'The biological significance of skin-to-skin contact and maternal odours', *Acta Paediatrica*, 2004, vol. 93, pp. 1560–2

'Neonatal recognition of the mother's face', *British Journal of Developmental Psychology*, 1989, 7 (1), pp. 3–15

'Newborn infants imitate adult facial gestures', *Child Development*, Jun. 1983, vol. 54 (3), pp. 702–9

'Is there an innate gaze module? Evidence from human neonates', *Infant Behavior and Development*, Feb. 2000, vol. 23 (2), pp. 223–9

'The perception of facial expressions in newborns', *European Journal of Developmental Psychology*, March 2007, vol. 4 (1), pp. 2–13

'Newborn infants perceive abstract numbers', *PNAS*, 23 June 2009, vol. 106 (25), pp. 10382–5

Introduction to Infant Development, Alan Slater and Michael Lewis (eds), 2nd edn, 2007, Oxford University Press

'A precursor of language acquisition in young infants', *Cognition*, 1988, vol. 29, pp. 143–78

'Prenatal experience and neonatal responsiveness to vocal expressions of emotion', *Developmental Psychobiology*, Nov. 1999, vol. 35 (3), pp. 204–14

'Newborns' cry melody is shaped by their native language', *Current Biology*, 15 Dec. 2009, vol. 19 (23), pp. 1994–7

'Can newborns discriminate between their own cry and the cry of another newborn infant?', *Developmental Psychology*, 1999, vol. 35 (2), pp. 418–26

How Babies Think, Gopnik, Alison, Meltzoff, Andrew and Kuhl, Patricia, Phoenix, 1999

'Newborn infants detect the beat in music', *Proceedings of the National Academy of Sciences*, 17 Feb. 2009, vol. 106 (7), pp. 2468–71

How much can newborns see?

'Development of contrast sensitivity in the human infant', *Vision Research*, 1990, vol. 30 (10), pp. 1475–86

'Does chromatic sensitivity develop more slowly than luminance sensitivity?', *Vision Research*, 1993, vol. 33 (17), pp. 2553–62

How do babies go from breathing nothing to breathing air?

Mayes' Midwifery, 13th edn, Baillière Tindall, 2004

'Regulation of breathing at birth', *Journal of Developmental Physiology*, Jan. 1991, vol. 15 (1), pp. 45–52

'Oxygen and reduced umbilical blood flow trigger the first breath of human neonates', *Acta Paediatrica*, 1992, vol. 34, pp. 660–2

'Neonatal transitional physiology: a new paradigm', *Journal of Perinatal and Neonatal Nursing*, 2002, 15 (4), pp. 56–75

Why do newborn babies smell so good?

'Olfaction and human kin recognition', *Genetica*, 1999, vol. 104, pp. 259–63

'Cortisol, hedonics, and maternal responsiveness in human mothers', *Hormones and Behavior*, Oct. 1997, vol. 32 (2), pp. 85–98

'Unique salience of maternal breast odors for newborn infants', *Neuroscience and Biobehavioral Reviews*, Jan. 1999, vol. 23 (3), pp. 439–49

Does skin-to-skin contact really soothe my baby?

'Early skin-to-skin contact for mothers and their healthy newborn infants', Moore E. R., Anderson, G. C. and Bergman, N., Cochrane Database of Systematic Reviews, 2007, issue 3, art. no.: CD003519

'Mother and newborn baby: mutual regulation of physiology and behavior – a selective review', *Developmental Psychobiology*, 2005, vol. 47 (3), pp. 217–29

'Postpartum maternal oxytocin release by newborns: effects of infant hand massage and sucking', *Birth*, Mar. 2001, vol. 28 (1), pp. 13–19

'Skin-to-skin care with the father after cesarean birth and its effect on newborn crying and prefeeding behavior', *Birth*, Jun. 2007, vol. 34 (2), pp. 105–14

How do newborns know to seek the nipple for food?

'Breast odour as the only maternal stimulus elicits crawling towards the odour source', *Acta Paediatrica*, Apr. 2001, vol. 90 (4), pp. 372–5

'The biological significance of skin-to-skin contact and maternal odours', *Acta Paediatrica*, Dec. 2004, vol. 93 (12), pp. 1560–62

'The secretion of areolar (Montgomery's) glands from lactating women elicits selective, unconditional responses in neonates', *PLoS ONE*, vol 4 (10), e7579

'An overlooked aspect of the human breast: areolar glands in relation with breastfeeding pattern, neonatal weight gain, and the dynamics of lactation', *Early Human Development*, Feb. 2012, vol. 88 (2), pp. 119–28

'Social chemosignals from breastfeeding women increase sexual motivation', *Hormones and Behavior*, Sept. 2004, vol. 46, pp. 362–70

What causes the Moro or 'startle' reflex?

'Infancy: Early Learning, Motor Skills, and Perceptual Capacity', *Child Development*, 7th edn, Berk, Laura E., Pearson/Allyn and Bacon, 2006

'The grasp reflex and Moro reflex in infants: hierarchy of primitive reflex responses', *International Journal of Paediatrics*, 2012, In Press, DOI: 10.1155/2012/191562

Baby Bodies

Are baby growth charts accurate?
'Using the new UK-WHO growth charts', *BMJ*, 2010, vol. 340, c1140
   'The WHO growth standards: strengths and limitations', *Current Opinion in Clinical Nutrition and Metabolic Care*, May 2012, vol. 15 (3), pp. 298–302

Do big babies grow into big adults?
'Birth weight and length as predictors for adult height', *American Journal of Epidemiology*, 1999, vol. 149 (8), pp. 726–9
   'Size at birth and gestational age as predictors of adult height and weight', *Epidemiology*, March 2005, 16 (2), pp. 175–81

Why don't babies have moles?
'Moles in children: what parents should know', American Academy of Dermatology fact sheet. http://www.skincarephysicians.com/skincancernet/moles_children.html
   'The effect of sun exposure in determining nevus density in UK adolescent twins', *Journal of Investigative Dermatology*, 2005, vol. 124, pp. 56–62

What causes colic?
'Managing infants who cry excessively in the first few months of life', *BMJ*, Dec. 2011, vol. 343, d7772
   'Guidelines for the diagnosis and management of cow's milk protein allergy in infants', *Archives of Disease in Childhood*, 2007, vol. 92, pp. 902–8
   'Infant colic and feeding difficulties', *Archives of Disease in Childhood*, 2004, vol. 89, pp. 908–12
   'A global, evidence-based consensus on the definition of gastroesophageal reflux disease in the pediatric population', *American Journal of Gastroenterology*, 2009, 104, pp. 1278–95
   'Focus on infantile colic', *Acta Paediatrica*, Sept. 2007, vol. 96 (9), pp. 1259–64
   'Patient information: acid reflux (gastroesophageal reflux) in infants (beyond the basics)', Winter, Harland S., Up to Date (website), Wolters Kluwer Health: http://www.uptodate.com/contents/acid-reflux-gastroesophageal-reflux-in-infants-beyond-the-basics

'Colic in babies: An NCT evidence-based briefing', *New Digest*, April 2008, vol. 38, pp. 23–8

What is the best way to settle a crying baby?
'Infant crying: a clinical conundrum', *Journal of Pediatric Health Care*, 2007, vol. 21, pp. 333–8

'Swaddling: a systematic review', *Pediatrics*, 1 Oct. 2007, vol. 120 (4), e1097–e1106

'Comparison of Behaviour Modification with and without swaddling as interventions for excessive crying', *Journal of Pediatrics*, Oct. 2006, vol. 149 (4), pp. 512–17, e2

'Chiropractic spinal manipulation for infant colic: a systematic review of randomised clinical trials', *International Journal of Clinical Practice*, Sept. 2009, vol. 63 (9), pp. 1351–3

'Distinguishing infant prolonged crying from sleep-waking problems', *Archives of Disease in Childhood*, April 2011, vol. 96 (4), pp. 340–4

'Managing infants who cry excessively in the first few months of life', *BMJ*, 2011, vol. 343, d7772

'Massage intervention for promoting mental and physical health in infants aged under six months', Underdown, A., Barlow, J., Chung, V. and Stewart-Brown, S., Cochrane Database of Systematic Reviews, 2006, issue 4, art. no.: CD005038

Why do some belly buttons become 'outies'?
'Umbilicus as a fitness signal in humans', *FASEB Journal*, 2009, vol. 23 (1), pp. 10–12

'In search of the ideal female umbilicus', *Plastic and Reconstructive Surgery*, vol. 105 (1), pp. 389–92

Do babies heal faster than adults?
'Age and pro-inflammatory cytokine production: wound-healing implications for scar-formation and the timing of genital surgery in boys', *Journal of Pediatric Urology*, June 2011, 7 (3), pp. 324–31

'Transient regenerative potential of the neonatal mouse heart', *Science*, 25 Feb. 2011, vol. 331 (6020), pp. 1078–80

Why do babies gnaw on things even before they start teething?
'Teething and tooth eruption in infants: a cohort study', *Pediatricians*, Dec. 2000, vol. 106 (6), pp. 1374–9

'Prospective longitudinal study of signs and symptoms associated with primary tooth eruption', *Pediatrics*, vol. 128 (3), Sept. 2011, e1–e6

Why is baby hair a different colour to adult hair, and why do babies lose their hair?

'Scalp hair characteristics in the newborn infant: scalp hair examination', *Advances in Neonatal Care*, 2003, vol. 3 (6), pp. 286–96

'Normal hair growth in children', *Pediatric Dermatology*, 1987, vol. 4 (3), n173–84

'Head and Neck Region', *Physical Diagnosis in Neonatology*, Fletcher, Mary Ann, Lippincott-Raven, 1998

Can newborns sweat?

'Sweating in preterm babies', *Journal of Pediatrics*, April 1982, vol. 100 (4), pp. 614–19

Are dummies good or bad for babies?

'Do pacifiers reduce the risk of sudden infant death syndrome? A meta-analysis', *Pediatrics*, Nov. 2005, vol. 116 (5), e716–23

'Risks and benefits of pacifiers', *American Family Physician*, 15 April 2009, vol. 79 (8), pp. 681–5

SLEEP

Can babies tell the difference between night and day?

'Developing circadian rhythmicity in infants', *Pediatrics*, 1 Aug. 2003, vol. 112 (2), pp. 373–81

Are a baby's sleep patterns inherited?

'Contribution of the photoperiod at birth to the association between season of birth and diurnal preference', *Neuroscience Letters*, 2006, vol. 406, pp. 113–16

'Genetic and environmental factors shape infant sleep patterns: a study of eighteen-month-old twins', *Pediatrics*, May 2011, vol. 127 (5), e1296–1302

'Individual differences, daily fluctuations, and developmental changes in amounts of infant waking, fussing, crying, feeding, and sleeping', *Child Development*, Oct. 1996, vol. 67 (5), pp. 2527–40

How can I get my baby to sleep through the night?
'What influences baby-sleeping behaviour at night? A review of evidence', *NCT New Digest*, April 2008, vol 42, pp. 25–30

'Help me make it through the night: behavioral entrainment of breast-fed infants' sleep patterns', *Pediatrics*, Feb. 1993, vol. 91 (2), pp. 436–44

'Use of a behavioural programme in the first three months to prevent infant crying and sleeping problems', *Journal of Paediatric Child Health*, June 2001, vol. 37 (3), pp. 289–97

Is co-sleeping good or bad for my baby?
'Parenting advice books about child sleep: co-sleeping and crying it out', *Sleep*, Dec. 2006, vol. 29 (12), pp. 1616–23

'Research overview: bed-sharing and co-sleeping', *NCT New Digest*, Oct. 2009, vol. 48, pp. 22–7

'Outcome correlates of parent–child bedsharing: an eighteen-year longitudinal study', *Journal of Developmental and Behavioral Pediatrics*, Aug. 2002, vol. 23 (4), pp. 244–53

'Parent–infant bed sharing and behavioural features in 2–4-month-old infants', *Early Child Development and Care*, 1999, vol. 149 (1) pp. 1–9

'Parenting practices in the Basque country: implications of infant and childhood sleeping location for personality development', *Ethos*, 1994, vol. 22, pp. 42–82

'Why babies should never sleep alone: a review of the co-sleeping controversy in relation to SIDS, bedsharing and breastfeeding', *Paedriatric Respiratory Reviews*, 2005, vol. 6, pp. 134–52

'Co-sleeping, an ancient practice: issues of the past and present, and possibilities for the future', *Sleep Medicine Reviews*, 2006, vol. 10, pp. 407–17

Will leaving my baby to cry cause any long-term damage?
'Behavioral treatment of bedtime problems and night wakings in infants and young children', *Sleep*, Oct 2006, vol. 29 (10), pp. 1263–76

'Parenting advice books about child sleep: cosleeping and crying it out', *Sleep*, Dec. 2006, vol. 29 (12), pp. 1616–23

'Asynchrony of mother–infant hypothalamic–pituitary–adrenal axis activity following extinction of infant crying responses induced during the transition to sleep', *Early Human Development*, Apr. 2012, vol. 88, pp. 227–32

'Five-Year Follow-up of Harms and Benefits of Behavioral Infant Sleep Intervention: Randomized Trial', *Pediatrics*, Sept. 2012, epub ahead of print, DOI: 10.1542/peds.2011–3467

What's more effective: introducing strict routines or immediately responding to my baby's demands?
'Infant crying and sleeping in London, Copenhagen and when parents adopt a "proximal" form of care', *Pediatrics*, 1 June 2006, vol. 117 (6), e1146–e1155

THE WHITE STUFF

How does breastfeeding work?
'Infant gastric physiology – ultrasound studies of the term and preterm infants', Seventh International Breastfeeding and Lactation Symposium, Vienna, Austria, 20–21 April 2012

Should I feed on demand or wait three to four hours between feeds?
'How breastfeeding works', *Journal of Midwifery and Women's Health*, Nov.–Dec. 2007, vol. 52 (6), pp. 564–70
   'Volume and frequency of breastfeedings and fat content of breast milk throughout the day', *Pediatrics*, 1 March 2006, vol. 117 (3), e387–95

Is breast really best for babies?
'"Breast is best": the evidence', *Early Human Development*, Nov. 2010, vol. 86, pp. 729–32
   'Breastfeeding and the use of human milk', *Pediatrics*, Feb. 2012, vol. 129, e827
   'Infant feeding matters', *Journal of Pediatrics*, vol. 159 (2), pp. 175–6
   'Children of the 90s: coming of age', *Nature*, 11 April 2012, vol. 484 (7393), pp. 155–8

How long do I need to breastfeed in order for my baby to reap the benefits?
'Breastfeeding and the use of human milk', *Pediatrics*, Feb. 2012, vol. 129, e827
   'Combination feeding of breast milk and formula: evidence for shorter breast-feeding duration from the National Health and Nutrition Examination Survey', *Journal of Pediatrics*, vol. 159 (2), pp. 186–90

Could combining breast- and formula-feeding offer babies the best of both worlds?

'Combination feeding of breast milk and formula: evidence for shorter breast-feeding duration from the National Health and Nutrition Examination Survey', *Journal of Pediatrics*, Aug. 2011, vol. 159 (2), pp. 186–90

What is in formula milk, and why does it taste fishy?

'Choosing a formula – what is the evidence for different milks and added ingredients?', *NCT New Digest*, July 2010, vol. 51, pp. 24–9

'Infant formula', *New Family Physician*, 1 April 2009, vol. 79 (7), pp. 565–70

'Infant formulas', *Pediatrics in Review*, vol. 32 (5), pp. 179–89

Does 'nipple confusion' really exist?

'Combination feeding of breast milk and formula: evidence for shorter breast-feeding duration from the National Health and Nutrition Examination Survey', *Journal of Pediatrics*, Aug. 2011, vol. 159 (2), pp. 186–90

'Which method of breastfeeding supplementation is best? The beliefs and practices of paediatricians and nurses', *Paediatric Child Health*, Sept. 2010, vol. 15 (7), pp. 427–31

'Cup feeding versus other forms of supplemental enteral feeding for newborn infants unable to fully breastfeed', Flint, A., New, K. and Davies, M. W., Cochrane Database of Systematic Reviews, 18 April 2007, issue 2, art. no.: CD005092

'Pacifiers and breastfeeding. A systematic review', *Archives of Pediatrics and Adolescent Medicine*, 2009, vol. 163 (4), pp. 378–82

'Pacifier use versus no pacifier use in breastfeeding term infants for increasing duration of breastfeeding', Jaafar, S. H., Jahanfar, S., Angolkar, M. and Ho, J. J., Cochrane Database of Systematic Reviews, 2011, issue 3, art. no.: CD007202

Does what I eat change the flavour of my milk?

'Differential transfer of dietary flavour compounds into human breast milk', *Physiology and Behaviour*, 3 Sept. 2008, vol. 95, issues 1–2, pp. 118–24

How much alcohol gets into breast milk?

'Ethanol and acetaldehyde in the milk and peripheral blood of lactating women after ethanol administration', *Journal of Obstetrics and Gynaecology of the British Commonwealth*, Jan. 1974, vol. 81. pp. 84–6

'Alcohol in breast milk', *Australia and New Zealand Journal of Obstetrics and Gynaecology*, Feb. 1985, vol. 25, pp. 71–3

'Effects of exposure to alcohol in mother's milk on infant sleep', *Pediatrics*, 1 May 1998, vol. 101 (5), e2

'Alcohol and breastfeeding. Dispelling the myths and promoting the evidence', *Nursing for Women's Health*, Dec.–Jan. 2011, vol. 14 (6), pp. 454–61

Is morning breast milk any different to evening breast milk?

'The possible role of human milk nucleotides as sleep inducers', *Nutritional Neuroscience*, Feb. 2009, vol. 12 (1), pp. 2–8

'Lactation is associated with an increase in slow-wave sleep in women', *Journal of Sleep Research*, Dec. 2002, vol. 11 (4), pp. 297–303

Can men lactate?

'Composition of breast fluid of a man with galactorrhea and hyperprolactinaemia', *Journal of Clinical Endocrinology and Metabolism*, 1 March 1981, vol. 52 (3), pp. 581–2

WEANING

When should I wean my baby on to solid foods?

'Trial on timing of introduction to solids and food type on infant growth', *Pediatrics*, Sept. 1998, vol. 102, pp. 569–73

'NCT evidence-based briefing. When can babies start on solids?', *New Digest*, Oct. 2002, pp. 22–5

'Does weaning influence growth and health up to eighteen months?', *Archives of Disease in Childhood*, Aug. 2004, vol. 89, pp. 728–33

'Paper for discussion: introduction of solid foods', UK Scientific Advisory Committee on Nutrition Subgroup on Maternal and Child Nutrition (SMCN), Agenda item 3, 29 Sept. 2003

'Timing of solid food introduction in relation to eczema, asthma, allergic rhinitis, and food and inhalant sensitization at the age of six years: results from the Prospective Birth Cohort Study LISA', *Pediatrics*, 1 Jan. 2008, vol. 121 (1), e44–52

'Timing of solid food introduction and risk of obesity in preschool-aged children', *Pediatrics*, 1 March 2011, vol. 127 (3), e544–e551

'Complementary feeding: a commentary by the ESPGHAN Committee on Nutrition', *Journal of Pediatric Gastroenterology and Nutrition*, Jan. 2008, vol. 46 (1), pp. 99–110

How do I get my baby to like vegetables?
'Variety is the spice of life: strategies for promoting fruit and vegetable acceptance during infancy', *Physiology and Behavior*, 22 April 2008, vol. 94 (1), pp. 29–38

'Decreasing dislike for sour and bitter in children and adults', *Appetite*, Jan. 2008, vol. 50 (1), pp. 139–45

'Increasing food familiarity without the tears. A role for visual exposure?', *Appetite*, Dec. 2011, vol. 57 (3), pp. 832–8

Is there any evidence for baby-led weaning being better than parent-led weaning?
'Baby knows best? The impact of weaning style on food preferences and body mass index in early childhood in a case-controlled sample', *BMJ Open*, 2012, vol. 2, e000298

'Is baby-led weaning feasible? When do babies first reach out for and eat finger foods?', *Maternal and Child Nutrition*, Jan. 2011, vol. 7 (1), pp. 27–33

Should I force my baby to eat?
'Maternal control of child feeding during the weaning period: differences between mothers following a baby-led or standard weaning approach', *Maternal and Child Health Journal*, 2011, vol. 15 (8), pp. 1265–71

'Does maternal control during feeding moderate early infant weight gain?', *Pediatrics*, 1 August 2006, vol. 118 (2), e293–e298

How many calories does a baby need?
'Energy requirements of infants', *Public Health Nutrition*, vol. 8 (7A), pp. 953–67

Should I avoid feeding my baby nuts or eggs to protect it against allergies?

'Complementary feeding: a commentary by the ESPGHAN Committee on Nutrition', *Journal of Pediatric Gastroenterology and Nutrition*, Jan. 2008, 46 (1), pp. 99–110

'Effects of early nutritional interventions on the development of atopic disease in infants and children: the role of maternal dietary restriction, breastfeeding, timing of introduction of complementary foods and hydrolyzed formulas', *Pediatrics*, 1 Jan. 2008, vol. 121 (1), pp. 183–91

THE BROWN STUFF

Are reusable nappies really greener than disposables?

'An updated lifecycle assessment study for disposable and reusable nappies', UK Environment Agency, Science Report, SC010018/SR2

'Disposable nappies for preventing napkin dermatitis in infants', Baer, E. L., Davies, M. W. and Easterbrook, K., Cochrane Database of Systematic Reviews, 2006, issue 3, art. no.: CD004262

Why does baby poo smell like mustard?

'Gas production by feces of infants', *Journal of Pediatric Gastroenterology and Nutrition*, May 2001, vol. 32, pp. 534–41

'My baby doesn't smell as bad as yours: The plasticity of disgust', *Evolution and Human Behavior*, Sept. 2006, vol. 27 (5), pp. 357–65

BABY BRAINS

Is the temperament of a newborn carried through into childhood and beyond?

'Temperament and personality: origins and outcomes', *Journal of Personality and Social Psychology*, 2000, vol. 78 (1), pp. 122–35

'Personality development: continuity and change over the life course', *Annual Review of Psychology*, Jan. 2010, vol. 61, pp. 517–42

'Behavioral genetics and child temperament', *Journal of Developmental Behavioral Pediatrics*, June 2005, vol. 26 (3), pp. 214–23

'Socio-emotional development: from infancy to young adulthood', *Scandinavian Journal of Psychology*, Dec. 2009, vol. 50 (6), pp. 592–601

'Does infancy matter? Predicting social behaviour from infant temperament', *Infant Behaviour and Development*, 1999, vol. 22 (4), pp. 445–55

When do babies develop a sense of themselves as individuals?
Introduction to Infant Development, Alan Slater and Michael Lewis (eds), 2nd edn, 2007, Oxford University Press

Can parents modify their baby's personality traits?
'Temperament and personality: origins and outcomes', *Journal of Personality and Social Psychology*, 2000, vol. 78 (1), pp. 122–35

'Personality development: continuity and change over the life course', *Annual Review of Psychology*, Jan. 2010, vol. 61, pp. 517–42

'The influence of temperament and mothering on attachment and exploration: an experimental manipulation of sensitive responsiveness among lower-class mothers with irritable infants', *Child Development*, Oct. 1994, vol. 65 (5), pp. 1457–77

'Reticent behavior and experiences in peer interactions in Chinese and Canadian children', *Developmental Psychology*, Jul. 2006, vol. 42 (4), pp. 656–65

Does the order of birth influence children's personalities?
'How birth order affects your personality', July 2010, *Scientific American Mind*

'Explaining the relation between birth order and intelligence', *Science*, 22 June 2007, vol. 316 (5832), p. 1717

'Why siblings are like Darwin's finches: birth order, sibling competition and adaptive divergence within the family', Frank Sullaway, *The Evolution of Personality and Individual Differences*, 2010, Oxford University Press

'Sources of human psychological differences: the Minnesota study of twins reared apart', *Science*, Oct. 1990, vol. 250 (4978), pp. 223–8

Why do twins develop different personalities?
'Neonatal temperament in monozygotic and dizygotic twin pairs', *Child Development*, Aug. 1990, vol. 61 (4), pp. 1230–7

Identically Different: Why You Can Change Your Genes, Spector, Tim, Weidenfeld & Nicolson, 2012

Do babies like some people better than others?
'Newborn infants prefer attractive faces', *Infant Behavior and Development*, vol. 21 (2), pp. 345–54

'Innate intersubjectivity: newborns' sensitivity to communication disturbance', *Developmental Psychology*, vol. 44 (6), pp. 1779–84

'Social evaluation by preverbal infants', *Nature*, vol. 450, pp. 557–9

How much do babies remember?
'The development of infant memory', *Current Directions in Psychological Science*, June 1999, vol. 8 (3), pp. 80–5

'What infant memory tells us about infantile amnesia: long-term recall and deferred imitation', *Journal of Experimental Child Psychology*, Jun. 1995, vol. 59, pp. 497–515

Why don't we remember being babies?
Introduction to Infant Development, Alan Slater and Michael Lewis (eds), 2nd edn, 2007, Oxford University Press

'Breaking the barrier? Children fail to translate their preverbal memories into language', *Psychological Science*, vol. 13 (3), pp. 225–31

Does nursery make babies sociable or stressed?
'Quality of care and temperament determine changes in cortisol concentrations over the day for young children in childcare', *Psychoneuroendocrinology*, Nov. 2000, vol. 25 (8), pp. 819–36

'Child care and the well-being of children', *Archives of Pediatric and Adolescent Medicine*, Jul. 2007, vol. 161 (7), pp. 669–76

'The NICHD study of early childcare and youth development. Findings for children up to age four and a half years', US Department of Health and Human Services

'Differential susceptibility to rearing experience: the case of childcare', *Journal of Child Psychology and Psychiatry*, Apr. 2009, vol. 50 (4), pp. 396–404

Are babies born in summer any different to winter babies?
'Born too late to win?' *Nature*, 21 July 1994, vol. 370, p. 186

'UK Football Association youth development proposals', 2011

'Relative age effect in youth soccer: analysis of the FIFA U17 World Cup competition', *Scandinavian Journal of Medicine and Science in Sports*, June 2010, vol. 20 (3), pp. 502–8

'Season of birth and onset of locomotion: theoretical and methodological implications', *Infant Behavior and Development*, Jan.–March 1993, vol. 16 (1), pp. 69–81

'Developmental vitamin D deficiency and risk of schizophrenia: a ten-year update', *Schizophrenia Bulletin*, Nov. 2010, vol. 36 (6), pp. 1073–8

'Born under a bad sign', *New Scientist*, 27 January 2007

Does being born at the start of the school year give you an academic advantage?

'Does when you are born matter? The impact of month of birth on children's cognitive and non-cognitive skills in England', a report to the Nuffield Foundation by Claire Crawford, Lorraine Dearden and Ellen Greaves (Institute for Fiscal Studies), Nov. 2011

LANGUAGE

When will my baby understand what I'm saying?

'At six–nine months, human infants know the meanings of many common nouns', Proceedings of the National Academy of Sciences, 28 Feb. 2012, vol. 109 (9), pp. 3253–8

Why do babies say 'Dada' before 'Mama'?

'Universal production patterns and ambient language influences in babbling: a cross-linguistic study of Korean- and English-learning infants', *Journal of Child Language*, March 2010, vol. 37 (2), pp. 293–318

'Some organization principles in early speech development', Green, Jordan R. and Ignatius S. B. Nip, *Speech Motor Control: New developments in basic and applied research*, Oxford Scholarship Online, 2012

When does baby babble take on meaning?

'Left hemisphere cerebral specialization for babies while babbling', *Science*, 30 Aug. 2002, 30, vol. 297 (5586), p. 1515

How Babies Think, Gopnik, Alison, Meltzoff, Andrew and Kuhl, Patricia, 2001, Phoenix

Do babies communicate without language?

'The value of vocalizing: five-month-old infants associate their own noncry vocalizations with responses from caregivers', *Child Development*, May–June 2009, vol. 80 (3), pp. 636–44

'Twelve-month-olds communicate helpfully and appropriately for knowledgeable and ignorant partners', *Cognition*, 2008, vol. 108, pp. 732–39

Do more physical babies develop language faster than sedentary babies?

'Learning to walk changes infants' social interactions', *Infant Behavior and Development*, Feb. 2011, vol. 34 (1), pp. 15–25

Why do women speak to babies in a silly voice?

'Cross-language analysis of phonetic units in language addressed to infants', *Science*, 1 Aug. 1997, vol. 277, pp. 684–6

'Effects of the acoustic properties of infant-directed speech on infant word recognition', *The Journal of the Acoustical Society of America*, July 2010, vol. 128 (1), pp. 389–400

'Fathers show modifications of infant-directed action similar to that of mothers', *Journal of Experimental Child Psychology*, 2012, vol. 111, pp. 367–78

When do babies start learning language?

How Babies Think, Gopnik, Alison, Meltzoff, Andrew and Kuhl, Patricia, 2001, Phoenix

'Infants show a facilitation effect for native language phonetic perception between six and twelve months', *Developmental Science*, March 2006, vol. 9 (2), F13–F21

'Foreign-language experience in infancy: Effects of short-term exposure and social interaction on phonetic learning', *Proceedings of the National Academy of Sciences*, 22 July 2003, vol. 100 (15), pp. 9096–9101

When is the best time to introduce a second language?

'Bilingual signed and spoken language acquisition from birth: implications for the mechanisms underlying early bilingual language acquisition', *Journal of Child Language*, 2001, vol. 28, pp. 453–96

'Early childhood bilingualism: perils and possibilities', *Journal of Applied Research on Learning*, April 2009, vol. 2, special issue, article 2

THE NEXT ONE

Am I likely to fall pregnant when breastfeeding?
'Risk of ovulation during lactation', *Lancet*, 1990, vol. 335 (8680), pp. 25–9

I had no problem getting pregnant last time, so why am I struggling to conceive now?
'Infertility for dummies', Perkins, Sharon and Meyers-Thompson, Jackie, 2007

'Risk of ovulation during lactation', *Lancet*, 1990, vol. 335 (8680), pp. 25–9

'Interactions between metabolic and reproductive functions in the resumption of postpartum fecundity', *American Journal of Human Biology*, 2009, vol. 21, pp. 559–66

Why do some women seem to give birth only to boys?
'Decline in sex ratio at birth after Kobe earthquake', *Human Reproduction*, Aug. 1998, vol. 13 (8), pp. 2321–2

'Exogenous shocks to the human sex ratio: the case of September 11, 2001 in New York City', *Human Reproduction*, Aug. 2006, vol. 21 (12), pp. 3127–31

'Declining sex ratio in a first nation community', *Environmental Health Perspectives*, Oct. 2005, vol. 113 (10), pp. 1295–8

'The sex ratio of children in relation to paternal preconceptional radiation dose: a study in Cumbria, northern England', *Journal of Epidemiology and Community Health*, 1996, vol. 50, pp. 645–52

'Trends in population sex ratios may be explained by changes in the frequencies of polymorphic alleles of a sex ratio gene', *Evolutionary Biology*, vol. 36 (2), pp. 190–200

'Offspring sex ratios at birth as markers of paternal endocrine disruption', *Environmental Research*, 2006, vol. 100, pp. 77–85

'Mother's occupation and sex ratio at birth', *BMC Public Health*, May 2010, vol. 10, p. 269

'Father's occupation and sex ratio of offspring', *Scandinavian Journal of Public Health*, 2007, vol. 35 (5), pp. 454–9

'Sex ratios, family size, and birth order', *American Journal of Epidemiology*, vol. 150 (9), pp. 957–62

'Beautiful British parents have more daughters', *Reproductive Sciences*, 2011, vol. 18 (4), pp. 353–8

Is there anything I can do to influence the gender of my baby?

'Experimental alteration of litter sex ratios in a mammal', *Proceedings of the Royal Society B*, Feb. 2008, vol. 275 (1632), pp. 323–7

'You are what your mother eats: evidence for maternal preconception diet influencing foetal sex in humans', *Proceedings of the Royal Society B: Biological Sciences*, Jul. 2008, vol. 275 (1643), pp. 1661–8

'Interpregnancy weight gain and the male-to-female sex ratio of the second pregnancy: a population-based cohort study', *Fertility and Sterility*, May 2008, vol. 89 (5), pp. 1240–4

'Maternal eating disorders influence sex ratio at birth', *Acta Obstetricia et Gynecologia Scandinavica*, 2008, vol. 87 (9), pp. 979–81

'Female gender pre-selection by maternal diet in combination with timing of sexual intercourse – a prospective study', *Reproductive Biomedicine Online*, Dec. 2010, vol. 21 (6), pp. 794–802

Do twins run in families?

'Twinning', *Lancet*, 2003, vol. 362, pp. 735–43

'A genome-wide linkage scan for dizygotic twinning in 525 families of mothers of dizygotic twins', *Human Reproduction*, 2010, vol. 25 (6), pp. 1569–80

'Time-lapse recordings reveal why IVF embryos are more likely to develop into twins', ESHRE Press Release, Payne, Dianna, 2007

Glossary

Adrenaline (epinephrine) – Hormone that regulates many of the body's functions including heart rate and blood pressure, and which plays a crucial role in the fight-or-flight response.

Amniotic fluid – Liquid that surrounds and cushions the developing baby in the uterus.

Amniotic sac – Sac containing the baby and amniotic fluid within the uterus. The amniotic sac is what ruptures when a woman's waters break.

Antenatal – The period before birth (pregnancy). Also known as prenatal.

Apgar score – A score ranging from one to ten that is used to assess the health and appearance of newborn babies at birth.

Attachment – The emotional bond between a baby and its caregiver, which psychologists believe affects the development of personality and the ability to form relationships throughout life.

Bilirubin – Yellow pigment produced during the natural breakdown of red blood cells. Helps give poo its brown colour and is responsible for the yellow discolouration of babies with jaundice.

Breech – A breech baby is one lying head-up in the uterus, rather than head-down. Babies lying in the breech position when labour begins are more likely to suffer complications during birth, so they are often delivered by C-section instead.

Caesarean section (C-section) – Surgical procedure in which an incision is made in a woman's abdomen in order to deliver one or more babies.

Catecholamines – A group of hormones that are released into the blood during times of physical or emotional stress.

Catheter – A thin plastic tube that is inserted into a vein or the bladder either to administer fluids or drain the bladder of urine without needing to go to the toilet.

Centers for Disease Control and Prevention – The United States' national public health institute.

Classical conditioning – A form of learning in which a stimulus such as food or pain gradually becomes associated with another stimulus.

Cochrane Collaboration – International non-profit organization that reviews the results of multiple clinical trials in order to help people make well-informed decisions about their health.

Colic – A condition in which an otherwise healthy baby cries for prolonged periods, without any obvious reason.

Collagen – Protein found in the skin and connective tissue that provides it with strength and firmness.

Controlled crying – An approach for teaching babies to sleep, which involves putting them to bed awake and leaving them for progressively longer periods if they cry, until they learn to fall asleep by themselves.

Cortisol – Hormone that controls many functions in the body, which is often released during times of stress.

Co-sleeping – Refers to sharing either a bed or a room with your baby.

Diastasis – Separation of the abdominal muscles that occurs during mid to late pregnancy.

Doppler probe – A hand-held device used to detect the baby's heartbeat during pregnancy.

Double-blind study – An experiment in which neither the researchers conducting the study, nor the volunteers taking part, know which treatment they are receiving.

Dream feed – Involves feeding your baby while it is asleep (usually between 10 p.m. and midnight) in the hope of getting a longer stretch of sleep for yourself.

Dummy – Also known as a soother or pacifier.

Embryo – A developing baby is referred to as an embryo until week ten of pregnancy, when it becomes a foetus.

Endometriosis – Medical condition in which cells from the lining of the uterus grow elsewhere in the body.

Endorphins – Also known as 'natural painkillers', these brain chemicals are released during pain and strenuous exercise and can promote feelings of well-being.

Epidural – During an epidural, a needle is injected into the small of your back and a small flexible tube inserted, which is then used to deliver painkilling drugs.

Episiotomy – A surgical cut made in the muscular tissue between the vagina and anus to help the baby get out.

Fibroids – Non-cancerous growths in and around the uterus, which can cause pain, heavy bleeding, and occasionally impact fertility.

Focal feed – An extra-large breast- or bottle-feed given between 10 p.m. and midnight to encourage babies to sleep for longer stretches during the night.

Foetus – A developing baby is referred to as a foetus from week ten of pregnancy until it is born.

Foetal alcohol syndrome (FAS) – Umbrella term for a range of alcohol-related birth defects.

Forceps – Surgical tongs that can be positioned around a baby's head during delivery and used to help ease it out.

Fore-milk – Milk released at the start of a breast feed, which is more watery, lower in fat and higher in carbohydrates than milk released towards the end of a feed.

Full-term – Babies are considered full-term from 37 weeks of pregnancy, and are called premature if they are born before this.

Habituation – A simple form of learning in which a person stops responding to a stimulus such as a sound or picture after being exposed to it lots of times.

Hind-milk – Creamy milk released towards the end of a breast feed.

Human chorionic gonadotrophin – Hormone produced from early pregnancy onwards, which can be detected in blood and urine and is what most pregnancy tests look for.

Induced labour – Labour is sometimes started artificially if a baby is overdue or if the mother or baby's health is thought to be at risk.

INSERM – A French research institution focused on public health.

Instrumental delivery – A birth in which instruments such as forceps or a vacuum device called a ventouse is used to help ease the baby's head out.

Jaundice – A common and usually harmless condition in newborn babies characterized by yellowing of the skin and the whites of the eyes. In rare cases jaundice can be a sign of an underlying health condition, or can lead to brain damage.

Lactose – A sugar found in milk.

Linea nigra – A brownish line extending from the belly-button down to below the panty-line that appears in most women during pregnancy and remains for about a year after giving birth.

Listeriosis – A serious infection caused by the bacterium *Listeria monocytogenes.*

Melanin – A brown pigment produced by the skin.

Menstruation – Monthly shedding of the uterine lining; also known as a period.

Meta-analysis – A way of combining the results of lots of different studies to get a more accurate picture of the risks or benefits of a treatment.

MRI scan – Magnetic resonance imaging (MRI) is a type of scan that uses strong magnets and radio waves to produce detailed images of the inside of the body.

Neural tube defect – Birth defects of the brain and spinal cord.

NHS – The UK's National Health Service.

NICE – The UK's National Institute for Health and Clinical Excellence which offers evidence-based guidance to doctors.

Nipple confusion – When a baby develops a preference for feeding from a bottle rather than the breast.

Oestrogen – The main female sex hormone, which is produced in large amounts during pregnancy and helps the uterus to grow, keeps the lining of the uterus nice and thick, boosts blood circulation and breast development, and also helps the baby to grow.

Ovulation – The monthly release of a mature egg by the ovaries.

Oxytocin – Hormone involved in triggering labour and breastfeeding, but which is also implicated in social bonding.

Peer review – A form of scientific self-regulation in which a study is critically read and reviewed by a researcher's peers before publication.

Pelvic-floor muscles – A group of muscles running between the pubic bone and tailbone that help support the bladder, vagina, uterus and bowels.

Perineal tear – A tear to the perineum that often occurs during childbirth. Tears are classified into four categories with a first-degree tear the most superficial, and a fourth-degree tear being the most serious, extending into the anus and rectum.

Perineum – The area of tissue between the vagina and anus.

Placebo – An inactive substance or mock treatment often used as a control in clinical studies to compare the effectiveness of the real treatment.

Placenta – Organ that connects a developing baby to its mother's uterus and allows the exchange of oxygen, nutrients and waste products between the two blood supplies.

Postnatal – The period of time after birth.

Pre-term – The birth of a baby before 37 weeks of pregnancy.

Progesterone – Pregnancy hormone that keeps the placenta working, keeps the lining of the uterus nice and thick, and stimulates breast growth.

Prolactin – Hormone that stimulates breast development and milk production.

Prostaglandins – Substances with many roles within the body, but which stimulate the cervix to soften and open, and the uterus to contract during birth.

Reflux – When the milk a baby has swallowed travels back up its food pipe and into its mouth.

Relaxin – Pregnancy hormone that causes the ligaments of the uterus and pelvis to loosen and relax to accommodate the ever-growing baby.

REM sleep – A normal stage of sleep characterized by rapid movements of the eyes under the eyelids. It is also when dreaming is thought to occur.

Serotonin – Chemical that regulates mood, appetite and sleep in the brain.

Sex ratio – The proportion of male to female births in a population.

SIDS – Sudden infant death syndrome, or cot death, is the sudden, unexpected and unexplained death of an apparently healthy baby.

Skin-to-skin contact – The practice of placing babies naked on their mothers' bare chests after birth.

Stages of labour – Labour is often divided into three stages: the first stage, when the uterus is contracting and the cervix opening; the second stage, when the baby is moving down the birth canal; and the third stage, when the placenta is delivered.

Swaddling – The practice of wrapping babies in a blanket or cloth in order to restrict the movement of their limbs.

Syntocinon – A synthetic form of oxytocin sometimes given to kick-start or speed up labour.

TENS – Transcutaneous electrical nerve stimulation involves the delivery of electrical impulses across the skin to block the transmission of pain signals.

Thyroid – The thyroid gland produces a number of hormones that control how quickly the body uses energy, makes proteins and responds to other hormones.

Trimester – Pregnancy is often divided into three trimesters, with the first trimester marking weeks one to twelve of pregnancy; the second trimester marking weeks 13 to 28, and the third trimester marking the final stage of pregnancy.

Ultrasound – An ultrasound scan uses sound waves to create images of the inside of the body.

Ventouse – A ventouse is a vacuum device which is applied to the baby's head to gently help pull it out of the mother's body.

Weaning – The introduction of solid foods to a baby's diet.

Acknowledgements

The original seeds of this book were planted at *New Scientist*, so I must thank my colleagues, in particular Richard Fisher, Rowan Hooper and Sumit Paul-Choudhury, for encouraging me to write and develop the 'Bumpology' column in the first place.

Thanks also to the great friends I made through the National Childbirth Trust: Natalie Moore, Janet Hanson, Rhona Cairns, Liz Foley, Hanna Gronborg and Michelle Ellis, for keeping me sane during the early baby days and for providing an ongoing source of questions and inspiration for this book. I am also indebted to Laura Gallagher, Nicola Jones and Clemmie Hooper, who gave up their time to read various drafts of the book and offered their feedback as mothers, editors and midwives.

I would also like to pay homage to my Facebook Bumpologists: Betony Bennett, Eloise Hansell, Amy Frary, Jo Marchant, Tim Boucher, Vicky Sumner, Joanna Hurley, Angelica Ronald and Joanna Hill – mums and dads who contributed their own questions about pregnancy and babies and provided a sounding board at various stages of the book's development.

I have picked many brains during the writing of *Bumpology*, and I am extremely grateful to the many researchers, doctors and other health workers – both quoted and unquoted – who patiently talked me through their research and unflinchingly responded to some of the quirkier questions I posed. I must give particular thanks to Philip Steer and Maria Elliott for sparing me so much of their time and expertise.

I'd also like to thank Susanna Wadeson and her colleagues at Bantam Press for their encouragement during the writing of this book, and Mark Henderson for telling Susanna about 'Bumpology' in the first place. Similarly, I must thank Patrick Barkham for introducing me to my excellent agent, Karolina Sutton at Curtis Brown, whose support has been invaluable.

Finally, thanks to my husband, Nic Fleming, for his patience and childcare wizardry, without whom this adventure in baby-making never would have been possible; and to our children, Matilda and Max, whose growth and development have provided me with so much inspiration and joy.

Linda Geddes is a London-based journalist who writes about biology, medicine and technology. Born in Cambridge, she graduated from Liverpool University with a first-class degree in Cell Biology. She has worked as both a news editor and reporter for *New Scientist* magazine, and has received numerous awards for her journalism, including the Association of British Science Writers' award for Best Investigative Journalism. She is married with two young children, Matilda and Max.